PREPARATIVE AND PROCESS-SCALE LIQUID CHROMATOGRAPHY

ELLIS HORWOOD SERIES IN
APPLIED SCIENCE AND INDUSTRIAL TECHNOLOGY

Series Editor: Dr D. H. SHARP, OBE, former General Secretary, Society of Chemical Industry; formerly General Secretary, Institution of Chemical Engineers; and former Technical Director, Confederation of British Industry.

This collection of books is designed to meet the needs of technologists already working in the fields to be covered, and for those new to the industries concerned. The series comprises valuable works of reference for scientists and engineers in many fields, with special usefulness to technologists and entrepreneurs in developing countries.

Students of chemical engineering, industrial and applied chemistry, and related fields, will also find these books of great use, with their emphasis on the practical technology as well as theory. The authors are highly qualified chemical engineers and industrial chemists with extensive experience, who write with the authority gained from their years in industry.

Published and in active publication

PRACTICAL USES OF DIAMONDS
A. BAKON, Research Centre of Geological Technique, Warsaw, and A. SZYMANSKI, Institute of Electronic Materials Technology, Warsaw
NATURAL GLASSES
V. BOUSKA *et al.*, Czechoslovak Society for Mineralogy & Geology, Czechoslovakia
POTTERY SCIENCE: Materials, Processes and Products
A. DINSDALE, lately Director of Research, British Ceramic Research Association
MATCHMAKING: Science, Technology and Manufacture
C. A. FINCH, Managing Director, Pentafin Associates, Chemical, Technical and Media Consultants, Stoke Mandeville, and S. RAMACHANDRAN, Senior Consultant, United Nations Industrial Development Organisation for the Match Industry
THE HOSPITAL LABORATORY: Strategy, Equipment, Management and Economics
T. B. HALES, Arrowe Park Hospital, Merseyside
OFFSHORE PETROLEUM TECHNOLOGY AND DRILLING EQUIPMENT
R. HOSIE, formerly of Robert Gordon's Institute of Technology, Aberdeen
MEASURING COLOUR: Second Edition
R. W. G. HUNT, Visiting Professor, The City University, London
MODERN APPLIED ENERGY CONSERVATION
Editor: K. JACQUES, University of Stirling, Scotland
CHARACTERIZATION OF FOSSIL FUEL LIQUIDS
D. W. JONES, University of Bristol
PAINT AND SURFACE COATINGS: Theory and Practice
Editor: R. LAMBOURNE, Technical Manager, INDCOLLAG (Industrial Colloid Advisory Group), Department of Physical Chemistry, University of Bristol
CROP PROTECTION CHEMICALS
B. G. LEVER, International Research and Development Planning Manager, ICI Agrochemicals
HANDBOOK OF MATERIALS HANDLING
Translated by R. G. T. LINDKVIST, MTG, Translation Editor: R. ROBINSON, Editor, *Materials Handling News*. Technical Editor: G. LUNDESJO, Rolatruc Limited
FERTILIZER TECHNOLOGY
G. C. LOWRISON, Consultant, Bradford
NON-WOVEN BONDED FABRICS
Editor: J. LUNENSCHLOSS, Institute of Textile Technology of the Rhenish-Westphalian Technical University, Aachen, and W. ALBRECHT, Wuppertal
REPROCESSING OF TYRES AND RUBBER WASTES: Recycling from the Rubber Products Industry
V. M. MAKAROV, Head of General Chemical Engineering, Labour Protection, and Nature Conservation Department, Yaroslavl Polytechnic Institute, USSR, and V. F. DROZDOVSKI, Head of the Rubber Reclaiming Laboratory, Research Institute of the Tyre Industry, Moscow, USSR
PROFIT BY QUALITY: The Essentials of Industrial Survival
P. W. MOIR, Consultant, West Sussex
EFFICIENT BEYOND IMAGINING: CIM and its Applications for Today's Industry
P. W. MOIR, Consultant, West Sussex
TRANSIENT SIMULATION METHODS FOR GAS NETWORKS
A. J. OSIADACZ, UMIST, Manchester

Series continued at back of book

PREPARATIVE AND PROCESS-SCALE LIQUID CHROMATOGRAPHY

Editor

G. SUBRAMANIAN Ph.D., MSc.
Department of Chemical Engineering
Loughborough University of Technology

ELLIS HORWOOD
NEW YORK LONDON TORONTO SYDNEY TOKYO SINGAPORE

First published in 1991 by
ELLIS HORWOOD LIMITED
Market Cross House, Cooper Street,
Chichester, West Sussex, PO19 1EB, England

A division of
Simon & Schuster International Group
A Paramount Communications Company

© Ellis Horwood Limited, 1991

All rights reserved. No part of this publication may be reproduced, stored in a retrieval system, or transmitted, in any form, or by any means, electronic, mechanical, photocopying, recording or otherwise, without the prior permission, in writing, of the publisher

Every effort has been made to trace all copyright holders, but if any have been inadvertently overlooked, the publishers will be pleased to make the necessary arrangements at the earliest opportunity.

Typeset in Times by Ellis Horwood Limited
Printed and bound in Great Britain
by Hartnolls, Bodmin, Cornwall

British Library Cataloguing-in-Publication Data

Preparative and process-scale liquid chromatography. —
(Ellis Horwood series in chemical engineering)
I. Subramanian, G. II. Series.
660
ISBN 0–13–678327–9

Library of Congress Cataloging-in-Publication Data

Preparative and process-scale liquid chromatography / editor, G. Subramanian.
p. cm. — (Ellis Horwood series in chemical engineering)
Includes bibliographical references and index.
ISBN 0–13–678327–9
1. Separation (Technology). 2. Liquid chromatography — Industrial applications.
I. Subramanian. G., 1935– . II. Series
TP156.S45P717   1991
660'.2842–dc20
91-29492
CIP

Table of contents

Preface 7

Chapter 1 — Comparison between analytical and preparative liquid chromatography 9
Neil R. Herbert
Jones Chromatography Ltd, Hengoed, UK.

Chapter 2 — Chromatography systems: design and control 39
A. F. Mann
European Technical Manager, Amicon Ltd, Stonehouse, UK

Chapter 3 — Technical structure of liquid chromatography separation plants: on the pilot and production scales .55
G. Munch
E. Merck, Darmstadt, Germany

Chapter 4 — The design of preparative chromatographic media 68
I. C. Chappell
Crosfield Chemicals, Warrington, UK

Chapter 5 — Column technology and packing materials in industrial preparative chromatography 82
Henri Colin
Prochrom, Champigneulles, France

Table of contents

Chapter 6 — The development of preparative and process chromatography . . .96
Derek A. Hill
The Wellcome Foundation Ltd, Dartford, Kent, UK

Chapter 7 — Determining operating parameters in process chromatography . 103
Michael Kelly
St. Nabor, France

Chapter 8 — Design and cost implications in process-scale chromatography . . 133
Kevin Connelly
ICI Pharmaceuticals, Macclesfield, UK

Chapter 9 — Factors of importance in the scale-up of ion-exchange process . . 146
Peter R. Levison
Whatman, Maidstone, UK

Chapter 10 — Continuous adsorption and chromatography in the purification of fermentation products 162
Gordon J. Rossiter
Advanced Separation Technologies Inc., Lakeland, USA

Chapter 11 — Applications of liquid chromatography to large-scale purification of bacterial proteins 224
Christopher R. Goward
PHLS Centre for Applied Microbiology and Research, Porton Down, UK

Chapter 12 — Protein purification: a new approach to affinity chromatography 236
Ken Jones
Affinity Chromatography Ltd, Ballasalla, Isle of Man

Chapter 13 — An intorduction to large-scale enantioseparation 250
Charles A. White
Fisons plc, Loughborough, UK

Chapter 14 — Preparative direct chromatographic separation of enantiomers on chiral stationary phases 265
J. N. Kinkel, K. Cabrera and F. Eisenbeiss
E. Merck, Germany

Index 284

Preface

Liquid phase chromatography may be defined as a means of separation of solutes by making use of their different adsorption and desorption characteristics when a solution containing a number of solutes is percolated through a column packed with a 'medium', which is usually a powdered or granular solid adsorbent. It was first developed as powerful analytical technique but is now being increasingly applied on the large scale too. Both analytical techniques and large-scale applications have been continuously developed by chemists and by chemical engineers respectively. The last two decades have seen a large number of published papers devoted to the many aspects of the technique, resulting in a better understanding of the basic principles involved.

The development of process-scale liquid chromatography has also been a powerful stimulus to instrument design and development. On the large scale the situation is now such that the technique can be applied routinely and with a high degree of reliability. Nevertheless there is still a great deal to be learned and much scope for process improvement.

It seemed timely to arrange an intensive short course to discuss the present state of the art and the interaction between analytical techniques and large-scale applications. The present book is the outcome. As the Table of contents will show, a wide range of topics has been covered. The papers have been published as they were presented, and are complete in themselves, so there is inevitably some repetition and overlap.

It is hoped that this broad coverage of this important subject will prove both informative and useful to all who are concerned with any aspects of preparative or process-scale liquid chromatographic separation.

1

Comparison between analytical and preparative liquid chromatography

N. R. Herbert
Jones Chromatography, Hengoed, UK

Analytical HPLC and preparative LC have many similarities but there are also many differences. Prepscale is obviously much larger with higher flow rates and a facility for the collection of fractions but there are other, perhaps less obvious differences. Many of these centre on the methodology employed to solve the problem. Preparative LC should be considered as a purification process. The problem is to purify the product of interest to the highest yield for the required purity as quickly, cheaply and easily as possible. Preparative LC will be used where simpler methods, e.g. hot filtration and recrystallization, would be unsuccessful or not feasible. It is the aim of this chapter to discuss the similarities and differences between analytical HPLC and preparative LC and assist the practitioner of the former to be skilled at the latter.

A rule-of-thumb definition for analytical, semi-preparative, preparative and process LC is given in Table 1.1. It is a matter of scale, and the size of the column or quantity to be collected is often used as a yardstick. It should be remembered however that preparative LC is any LC where fractions are collected. A full production run of a complex peptide may only yield 100 mg whereas a research worker may prepare 1 g of a simple compound to do further studies before rejection of the product.

A brief summary of the operational similarities and differences between analytical and preparative liquid chromatography is given in Table 1.2. It can be seen that there are more differences than similarities which is in keeping with the different aims of the two techniques. The mobile phase is more restricted in the preparative case for reasons of cost, toxicity and product work-up rather than any chromatographic reason. The flow rate is higher than optimum in the preparative case in order to increase throughput. In the analytical case the use of optimum flow rate leads to improved resolution. The system back pressure is only a problem in analytical HPLC when using long columns of 3 μm materials with viscous solvents, or if premature wear of the pump seals leads to maintenance problems. In the preparative case there

Table 1.1 — Definitions

	Column i.d.	Column typical dimensions
Analytical	1 mm–6 mm	250 mm × 4.6 mm i.d.
Semi-prep	7 mm–25 mm	250 mm × 10 mm/22.5 mm i.d.
Prep	25 mm–150 mm	300 mm × 50 mm/100 mm i.d.
Process	> 150 mm	1000 mm × 300 mm i.d.

	Quantities of product
Analytical	µg
Semi-prep	mg
Prep	g
Process	kg

Table 1.2 — Operational conditions

Condition	Analytical	Preparative
Mobile phase	Wide range	More restricted
Flow rate	Often theoretical optimum	Often higher than theoretical optimum
Back pressure	Unimportant within system constraints	Important
Speed of run	Important (short)	Important (longer)
Sample load	Non-overload	Overload
Sample volume	Small	Large
Temperature control	Sometimes desirable	Less common
Sample stability	Important	Very important

is usually less of a buffer of back-pressure potential and preparative systems are often worked harder. Temperature control is a useful accessory in analytical HPLC especially where multisite operation of an assay is common. It is less common in preparative LC for engineering reasons and where site-to-site reproducibility is less important, even though the final product specification is identical.

The features of the equipment used in the two techniques are different and Table 1.3 gives a comparison of analytical and preparative systems. Included under the umbrella 'Preparative' are semi-preparative, preparative and process systems. A

Table 1.3 — Equipment features

Feature	Analytical	Preparative
Range of equipment	Many suppliers	More restricted
Specification of equipment	Similar	Wide range
Cost of equipment (pump + detector + sample injection system + recorder)	Similar e.g. £7500	Wide range, dependent on specification e.g. £8000–£250 000
System ancillaries	Minimal	0–£500 000
Explosion proof	No	Dependent on scale of operation and operating conditions
System weight	Light	Heavy
Auto/manual operation	Sometimes important	Auto important
Use requires investigation of safety/government/company regulations	Not usually	Often

much wider specification range of equipment is available for preparative LC compared to analytical HPLC. When choosing equipment it is very important to decide at what scale one is working and choose a system accordingly; under-specified equipment is very restrictive but over-specified systems are expensive. Equipment use, e.g. automatic or manual, must be considered in advance since a manual system may not be easily upgradeable to fully automatic operation. In analytical systems, system ancillaries are minimal but in preparative systems and especially at the process scale, they are very important. Examples of ancillaries are rotary evaporators or freeze-driers, solvent recycling/purification systems, explosion proofing, additional fume hoods or even a new building! The scale of use of toxic or flammable solvents often requires an investigation of company or government safety regulations.

In Table 1.4 the parameters or features of the components of the equipment used in analytical and preparative LC are compared. The differences are due to the differing requirements of the two techniques. Preparative LC should not be considered as simply a large LC system with a large column but as a means of separating the product of interest from everything else. Reproducibility of pump flow rate and gradient or isocratic mixing is very important in the preparative case as well as analytically. UV or refractive index (RI) detectors typically do not require high sensitivity for preparative work. Indeed the flow cells are designed to reduce sensitivity and so not overload the detector when high concentrations of sample are

Table 1.4 — System parameters

Parameter	Analytical	Preparative
Pump		
Pulse free	Very important	Important
Flow rate reproducibility	Very important	Very important
Flow rate accuracy	Very important	Important
Flow rate specification	0.1–9.9 ml/min	Varied, e.g. 1–150 ml/min 25–1000 ml/min
Pressure rating	6000 psi typically	Varied, e.g. 250–5000 psi often dependent on flow rate
Gradient capability	Both gradient + isocratic systems	Usual
Gradient mixer		
reproducibility	Very important	Very important
Detector		
UV-type	Fixed/variable	Variable usual
Sensitivity	High sensitivity	Lower sensitivity
Cell path length	1 cm typical	< 1 mm typical
Flow capability	< 10 ml/min	< litres/min
Pressure rating	Often 200–400 psi	High, > 2000 psi
Cell number	1	1 or 2
Injection	6-port valve/ autosampler typically	6-/10-port valve, through-pump valve, second pump
Quantitation	Very important	Less important
Data storage	Important	Important
Fraction collector	Not important	Very important
Column system		
Type	Compression fittings Analytical cartridge system	Flange fittings Preparative cartridge systems, Packer/column combinations
Column packing	Slurry packed self/manufacturer	Slurry/axial compression/radial compression/dry self/manufacturer
Packing materials	Spherical/irregular Performance/ selectivity choice	Spherical/irregular Performance/ selectivity/ loadability choice
	Analytical grade, e.g. Spherical 5 μm ODS, £9500/kg	Preparative grade, e.g. Spherical 15 μm ODS, £2000/kg Irregular 15–20 μm Silica, £250/kg.

passed through them. The RI detector will often have sufficient sensitivity and does not require a chromophore to function. It cannot, however, be used at the present time with gradient elution. A must for preparative detectors is high flow rate and pressure capability. The use of stream splitters before the detector cell is not recommended; a cell capable of taking full flow and pressure is much to be preferred.

Injection techniques differ and will be considered in more detail later. Quantitation is less important in preparative chromatography but data storage is very important for GMP reasons. As might be imagined, preparative and analytical column systems are different. Compression fittings reach the limit of their usefulness at 1″ i.d. level. For 1″ and above, columns of a flange-fitting design are much more usual although some companies use a cartridge design for extra economy. Columns may be user- or manufacturer-packed and whilst slurry packing is almost universal for analytical columns, preparative columns may be slurry packed, dry packed, packed by radial compression or packed by axial compression. The latter is currently the most popular for particles which are too small to be dry packed (i.e. less than 40 μm).

The criteria for packing material choice is similar in the two techniques, namely performance and selectivity, but in the preparative systems cost and loadability also need to be considered. At this stage the appearance of the chromatogram may be mentioned. Everyone is well used to the textbook analytical HPLC chromatograms which consist of sharp, symmetrical, well-resolved peaks. In preparative LC the ideal appearance of the chromatogram will be grossly overloaded peaks, probably with poor shape and with just enough resolution to obtain the specified yield, purity and throughput criteria. Appearances can be deceptive!

SYSTEM OPERATION

In a typical analytical HPLC system a fixed flow rate, gradient or isocratic elution and manual valve or auto-injection are used, the results being quantified using a data system.

In a typical preparative run the above methodology may be used with the addition of a fraction collector. However, to speed up or optimize the run, additional techniques or alternative approaches may be used; e.g.

— backflush
— sample pre-concentration
— recycle
— multi-column operation
— adsorption/desorption techniques
— frontal elution/displacement chromatography
— flash chromatography
— combinations of above techniques
— special techniques, e.g. flip-flop chromatography described by H. Colin, Prochrom, France [1].

In addition, the possibility of reworking impure fractions should be considered.

These techniques will be returned to later after considering the following operational differences.

Sample preparation

Analytical: dissolve in a suitable solvent, preferably the mobile phase.

Preparative: as above but, if possible, use a weaker solvent than the mobile phase. Circumstances will tend to dictate a solvent but remember that much larger volumes of sample solvent will typically be used. It may be necessary to modify the operational conditions to take into account the constraints imposed by the sample solvent. Filter before use.

Sample stability

Analytical: this is important for accurate quantitation only. The sample must be stable for the time of analysis, including sample preparation time.

Preparative: this is very important for purity and yield of product. The product of interest should be checked for stability in the sample solvent, in the mobile phase and on the column matrix. If there is a problem and alternative solvents or columns cannot be used consider reducing the time of the component on the column, or elute into a 'friendly' solvent, e.g. elution into an aqueous buffer of proteins in reversed-phase separations.

Product purity and yield

Analytical: fully resolved peaks with prior test of single peak for purity.

Preparative: a known injection volume from a known sample volume is tested analytically for peak homogeneity. The use of two analytical systems, e.g. silica/ODS, to check purity and yield is beneficial. Here selectivity differences between hydrophilic and hydrophobic phases will tend to differentiate impurities. Also, slow-running impurities of breakdown products will tend to elute faster than the peak of interest in the other column.

Obviously 100% yield and 100% purity are ideal but are they affordable? The yield should be maximized to give the required purity 90% +, 95% +, 98% +, etc. The remaining impurities may be removed much more easily at a later stage in the purification process. Alternatively, the impure fractions may be pooled and rechromatographed later to improve the yield.

Injection techniques

In analytical HPLC the use of a 6-port valve, e.g. Rheodyne or Valco, is almost universal for manual injection with auto-injection and some valve switching techniques used in the automatic case.

In preparative LC a 6- or 10-port valve may be used for injections up to about 10 ml. For injection volumes above this a second pump may be used to load sample directly onto the column or a through-pump injection system may be used. The second pump option can be used to load any volume but the through-pump valve has a practical minimum of around 10 ml. Both these techniques are easy to operate, especially the latter, and are capable of good results.

Economic factors
In analytical HPLC the initial cost of the equipment and the cost of the columns or packing materials are usually most important, especially the former.

In preparative or process LC the initial cost of the equipment and the column or packing material are usually not very significant when the cost is translated into contribution to the overall cost of purification. The major cost areas are often solvent and ancillary equipment and, in non-automated systems, labour. It is important to cost all the areas relevant to the total cost of the purification and then to see if the major ones can be easily reduced. Cost savings on say, packing material, may be insignificant compared to a large reduction in labour costs brought about by automated operation [2]. Table 1.5 gives a listing of parameters to be considered.

Scale-up
All preparative separations start with an analytical run and, in fact, much of the ground work is done at this scale. It is very important and should not be omitted. When scaling-up, it is important to consider several important parameters, which will aid in the transition from analytical HPLC to preparative LC.

Number of theoretical plates
We require to calculate the number of theoretical plates for a given separation in the analytical case and the preparative case. Preparative columns are traditionally operated at well above optimum flow rate for the packing and overloaded with sample in order to increase throughput. Both these factors reduce column efficiency and peak resolution. However, if loading and flow rate are increased too much then the separation is lost and peak purity is reduced to an unacceptable level. It has been shown that a ratio of $N_{anal.}$ to $N_{prep.}$ of around 3 is often optimum [1]. The use of a scout column — an analytical-sized column packed with preparative material — one-third as long as the preparative column, is a convenient aid to attaining this criteria in method development.

For Gaussian peaks a resolution of $R_s = 1.5$ gives good baseline resolution. For a typical average preparative k' value of 5 and a shorter one of 2, the number of plates needed per column for various values is given in Table 1.6. It can be seen that the higher the α selectivity value the easier the separation and hence the higher the level of potential purity. For a robust method it is best to work with as high an α value as possible within the constraints of cycle time. Time spent on optimizing the analytical chromatography to increase α is time well spent.

Column length and particle size for a given plate number
If we assume a moderately well-packed column ($h = 3$) then Table 1.7 gives various values for column length (L) and mean particle diameter (d_p) for values of $N_{anal.}$ and $N_{prep.}$ required for the separation.

Table 1.5 — Economic considerations

System
— Cost of preparative system
— Lifetime of system
— Running costs and maintenance costs

These lead to a total equipment cost per annum: (E)

Columns
— Cost of columns or cartridge holder/cartridges
— Lifetime of columns and guard columns
— Number of columns and guard columns used per year

These lead to a total column cost per annum: (C)

Chromatography running costs
— Quantity of product required to be processed per year to produce the specified yield and purity
— Quantity of product capable of being processed per run
— Number and duration of runs per day (or week, year, etc.)
— Volume of solvent used per run and per year
— Cost of new solvent
— Cost of waste disposal, solvent recycling/purification
— Cost of solvent analysis per year
— Cost of power per year

These lead to total running costs per annum: (S)

Labour
— The number of runs required per year and the amount of operator time/technician time/chemist/engineering/manager time per year for the process can be determined
— Salaries for the above and the percentage of time spent on the process per year can be determined

These lead to total labour costs per annum: (L)

Ancillary costs
— Product work-up
— In process QC
— etc.

These lead to total ancillary cost: (A)

$$\text{Cost of LC purification process} = \frac{E + C + S + L + A}{\text{Quantity of product}} \text{ per kilo per annum}$$

Small changes in the yield and purity criteria can have a large effect on the £/kg cost of LC purification.

Table 1.6 — Required column efficiency with variation in relative retention

R_s	k'_{av}	α	N	R_s	k'_{av}	α	N
1.5	2	2	81	1.5	5	2	52
1.5	2	1.5	324	1.5	5	1.5	207
1.5	2	1.3	900	1.5	5	1.3	576
1.5	2	1.2	2025	1.5	5	1.2	1296
1.5	2	1.1	8100	1.5	5	1.1	5184
1.5	2	1.05	34200	1.5	5	1.05	20736
1.5	2	1.03	90000	1.5	5	1.03	57600
1.5	2	1.02	202500	1.5	5	1.02	129600
1.5	2	1.01	810000	1.5	5	1.01	518400

$$R_s = \text{resolution}, \quad R_s = \frac{2(t_2 - t_1)}{(w_1 + w_2)} = \frac{1}{4}(\alpha - 1)\sqrt{N}\left(\frac{k'_{av}}{k'_{av} + 1}\right)$$

k'_{av} = average capacity factor, $k'_a = \dfrac{t_a - t_o}{t_o}$

α = relative retention value, i.e. $\alpha = \dfrac{k'_a}{k'_b}$

N = column efficiency, $N = 5.54\left(\dfrac{t}{w_{1/2}}\right)^2$

expressed as theoroetical plates per column, ($w_{1/2}$ is the peak width at half height).

As a guide, a very well packed 5 µm analytical column, 25 cm long, would have an N value of 20 000 plates per column.

A k' value of 5 is more typical in preparative runs

Pressure requirements for the above L/d_p options

Table 1.8 gives pressure drop values (ΔP) for a reversed-phase example at constant flow varying L with d_p to obtain a constant value for N. For irregular particles the pressure values are approximately 1.5–2 times those given above for spherical packings. However, this is only an approximation since back pressure depends on the particle size distribution and especially on the presence of fine particles.

Flow rate vs. particle size

Small particles produce their optimum efficiency at higher flow rates than large particles. Table 1.9 gives data on optimum flow rate for various-sized particles in a given column configuration. It also gives the optimum flow rate for various-sized columns at constant linear flow velocity. The relationship between efficiency and flow rate has been described by Knox and others.

18 Comparison between analytical and preparative liquid chromatography [Ch. 1

Table 1.7 — For constant column efficiency (N), variation of column length (L) and particle size (d_p)

$N_{anal.} = 6000$ $N_{prep.} = 2000$ $h = 3$	L (mm)	d_p (µm)	d_p (µm)	L (mm)
	2000	111	50	900
	1000	55	20	360
	500	28	15	270
	250	14	12	216
	150	8.3	10	180
	100	5.6	8	144
			5	90
			4	72
			3	54
			1	18

$N_{anal.}$ = efficiency for an analytical load level (plates per column)
$N_{prep.}$ = efficiency for a preparative load level (plates per column)
h = reduced plate height

$$N = 5.54 \left(\frac{t}{w^{1/2}} \right)^2$$

$$N = \frac{L}{h} \times \frac{1}{d_p} \times 10^3 \text{ plates per column}$$

$h = H/d_p$ and $H = L/N$
H = height equivalent of a theoretical plate (mn)
d_p = particle diameter (in microns)
Assume $N_{prep.} = \frac{1}{3} N_{anal.}$ for the solute of importance.

Elution time vs. particle size

The optimum flow-rate considerations given in Table 1.9 allow one to calculate the relative elution times for a given run. This can be correlated directly with throughput in the preparative application. The values are given in Table 1.10 which combines the data from Table 1.9 to give values for d_p, flow rate and comparative run times for the columns of constant efficiency (N).

Loading vs. column size

In analytical separations the relationship between the concentration of sample in the stationary and mobile phases is constant and independent of sample size. The relative ratio of solute in the two phases is the basis of chromatographic separations and the relationship manifests itself in varying degrees of retention for the solutes. In the case of preparative separations the situation is more complex and the retention varies as a function of sample size. Thus k' varies and the peak shape becomes

Comparison between analytical and preparative liquid chromatography

Table 1.8 — Influence of particle size on column back pressure

L = length of column to attain above plate number (mm)
u = optimum linear velocity of mobile phase down the column (mm/s)
Pressure in bar and psi for linear velocities (u (mmls) and $10 \times u$ (mmls)). For flow rates (ml/min) see Table 1.9

d_p (μm)	L (mm)	u (mm/s)	ΔP (bar)	ΔP (psi)	$10u$ (mm/s)	ΔP (bar)	ΔP (psi)
1	18	3.50	134	1948	35	1340	19480
3	54	1.17	44.7	650	11.7	447	6500
4	72	0.88	33.4	486	8.8	334	4860
5	90	0.70	26.8	390	7.0	268	3900
8	144	0.44	16.75	244	4.4	167.5	2440
10	180	0.35	13.40	195	3.5	134	1950
12	216	0.29	11.16	162	2.9	111.6	1620
15	270	0.23	8.93	130	2.3	89.3	1300
20	360	0.18	6.70	97	1.8	67	970
50	900	0.07	2.66	39	0.7	26.6	390

Mobile phase = methanol/water (65/35, v/v)
Temperature = ambient (20°C)
$N_{prep.}$ = 2000 plates (equivalent to $N_{anal.}$ = 6000)
The use of a less viscous mobile phase will lower the column operating pressure.

$$\text{Pressure } (P) = \frac{\phi \eta L F_c}{\pi/4 \, d_o^2 d_p^2 E} = \frac{\phi \eta L}{t_o d_p} \text{ (bar)}$$

$$= \frac{9.0 \times 10^6 \, \eta \, (\text{mNs/m}^2) \, L \, (\text{m}) \, F_c \, (\text{ml/min})}{d_o^2 \, (\text{mm}) \, d_p^2 \, (\mu m)}$$

where ϕ = flow resistance parameter (assumed to be 500)
 η = mobile phase viscosity (cP or mNs/m²)
 L = column length (m)
 F_c = flow rate (ml/min)
 d_c = column i.d. (mm)
 d_p = particle diameter (μm)
 E = porosity (assumed to be 0.7 for porous spheres)

The pressure is independent of the column diameter if constant linear velocities are used.

Linear velocity, $u = \dfrac{L}{t_o}$ (mm/s)

where t_o = time for elution of unretained peak

Table 1.9 — Influence of particle size on flow rate

d_p (μm)	u (mm/s)	Flow rate (ml/min) for the following column i.d.s (mm)				
		4.60	10.00	22.50	50.00	100.00
1	3.50	3.49	16.50	83.51	412.39	1649.55
3	1.17	1.16	5.50	27.84	137.46	549.85
4	0.88	0.87	4.12	20.88	103.10	412.39
5	0.70	0.70	3.30	16.70	82.48	329.91
8	0.44	0.44	2.06	10.44	51.55	206.19
10	0.35	0.35	1.65	8.35	41.24	164.96
12	0.29	0.29	1.37	6.96	34.37	137.46
15	0.23	0.23	1.10	5.57	27.49	109.97
20	0.18	0.17	0.82	4.18	20.62	82.48
50	0.07	0.07	0.33	1.67	8.25	32.99

Linear flow velocity, $u = \dfrac{v \cdot D_m}{d_p} = \dfrac{L \text{(mm)}}{t_o \text{ (s)}}$

v = reduced velocity; let this equal 5 for optimum practical performance.
D_m = diffusity coefficient. In the reversed-phase example above,
(65:35 MeOH:H$_2$O) $D_m = 0.7 \times 10^{-9}$ m^2/s

D_m is a function of mobile-phase viscosity and is usually quoted by the Wilke-Chang equation.

$$D_m = 7.4 \times 10^{-12} \sqrt{\psi M_{solv.}} \cdot \dfrac{T}{\eta} \cdot V_{solute}^{0.6} \text{ (m}^2\text{/s)}$$

where
ψ = solvents association factor (1 for non-polar solvents, 1.9 for methanol, 1.5 for ethanol, 2.6 for water)
η = viscosity
$M_{solv.}$ = solvent molecular weight (g)
T = temperature (k)
V_{solute} = molar volume (ml), i.e. molecular weight of solute/density.

Variation in the value of D_m directly affects the optimum flow rate since d_p and v are constant. Examples of the variation in D_m are given in Table 1.14.

Table 1.10 — Variation of run time with particle size and column dimensions

Optimum linear velocity and flow rate are used for each particle size and the length of column is chosen to give a constant value for N using $v = 5$. An $N_{prep.}$ value of 2000 is used again.
Run time is asumed to be $5 \times t_o$.

d_p (μm)	u (mm/s)	All column diameters L (mm)	t_o (min)	Run time ($5t_o$) (min)
1	3.50	18	0.09	0.45
3	1.17	54	0.77	3.85
4	0.88	72	1.36	6.80
5	0.70	90	2.14	10.70
8	0.44	144	5.45	27.25
10	0.35	180	8.57	42.85
12	0.29	216	12.41	62.05
15	0.23	270	19.57	97.85
20	0.18	360	33.30	166.50
50	0.07	900	214.30	1071.50

Linear flow velocities (u) are independent of column i.d. $v = nu$; Reduced mobile-phase velocity

distorted, changing from Gaussian to triangular. This is shown in Fig. 1.1. It can be seen that the break-through point of the peak and the peak maxima is earlier. The point of the triangle is near the k' value for the analytical system. Pure volume overload peaks are flat-topped as would be expected since the sample is loaded over a longer time period and volume of column.

If the sample is only sparingly soluble in the mobile phase a peak which fronts rather than tails is produced and the k' values are increased.

The load capability of a column will vary with the type of material in the column and is a function of the surface area of the packing, amount of material in the column, carbon loading of the material, etc. For silica-based materials a typical rule-of-thumb value is 1 mg per g of packing before overloading occurs. In preparative LC, however, the column is typically overloaded to as high a level as possible to increase throughput. Typical column sizes, their material content and 'typical' load values are given in Table 1.11. For maximum loading the sample should be dissolved if possible in a solvent weaker than the mobile phase.

The loadability will have to be verified in the actual application using the preparative packing material. The use of a scout column is useful for this task. The load for larger columns can be scaled-up using the relative material weights as scale-

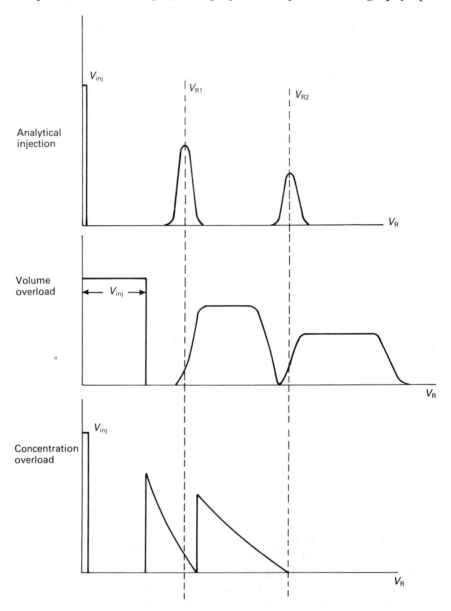

Fig. 1.1 — Overload. Analytical injection. Volume overload and concentration overload. (Ref. [2].)

up factors. The actual weight of material in the column will vary with the true density of the porous particle which in turn is a function of the pore volume of the material.

Associated with loadability is the actual injection volume. Since preparative systems work under overload conditions one can either 'volume overload', 'concentration overload' or a combination of the two. There is evidence to suggest that

Table 1.11 — Variation of column size, weight of packing and load capacity

L (mm)	d_c (mm)	Column volume (litre)	Packing weight (kg)	Load 1 mg of sample per gram	Load 10 mg of sample per gram	Scale-up factor
250	4.6	0.0032	0.0032	0.032	0.032	1
250	10	0.02	0.02	0.02	0.15	5
250	22.5	0.10	0.08	0.08	0.77	24
500	22.5	0.20	0.15	0.15	1.53	48
250	50	0.49	0.38	0.38	3.78	96
500	50	0.98	0.76	0.76	7.56	192
500	100	3.93	3.02	3.02	30.24	766
1000	100	7.86	6.05	6.05	60.48	1531
1000	180	31.42	24.19	24.19	241.93	6125
2000	180	62.84	48.39	48.39	483.87	12250
1000	360	125.68	96.77	96.77	967.74	24500
2000	360	251.36	193.55	193.55	1935.47	49000

Assume packing density is 0.77 g/cm^3. This value will vary with the pore volume of the packing material.

'concentration overload' is preferable but that up to a certain limiting value, volume of sample does not adversely affect the chromatography. Beyond this value, peak broadening occurs. Typically the value of this limiting value is 20–50% of the volume of the peak arising from a small injection volume. In practice the concentration of this volume of sample is increased until peak resolution deteriorates to the lowest acceptable value. Table 1.12 gives the variation in peak volumes with column size.

To optimize injection, a uniform distribution of sample over the column cross-section is required. The overload is thus uniform over the column cross-section. This is easier to achieve with a larger injection volume.

The value of the peak volume will increase with the degree of retention on the column and so at higher k' values a higher volume can be loaded. Where solubility is a problem longer peak retention may be required to obtain the required load per run. However, run time will be increased so throughput per hour may or may not be improved.

The loss of column efficiency owing to overloading of sample is generally higher with columns packed with smaller particle size materials than ones containing larger particles.

So far we have considered a single, ideal solute in the column. In practice this is never the case and allowance should be made for interactions beween the solutes. Prediction is not easy and for separations with large α values the effects should be

Table 1.12 — Variation of peak volume with column size

Column (mm × mm i.d.)	Flow rate (ml/min)	t_o (min)	Peak volume (V_r/ml)	Typical maximum preparative injection volume, 50% of the peak volume (ml)
250 × 4.6	0.70	5.95	2.2	1.1
250 × 10	3.31	5.95	10.6	5.3
250 × 22.5	16.75	5.95	53.5	26.6
500 × 22.5	16.75	11.90	107.0	53.5
250 × 50	66.99	5.95	214.0	107.0
500 × 50	66.99	11.90	428.1	214.1
500 × 100	267.96	11.90	1712.5	856.2
1000 × 100	267.96	23.81	3425.0	1712.5
1000 × 200	1071.83	23.81	13699.9	6850.0
2000 × 200	1071.83	47.62	27399.8	13700.0
1000 × 400	4287.33	23.81	54799.7	27400.0
2000 × 400	4287.33	47.62	109599.3	54799.7

$t_o = \dfrac{L}{u}$. In this example $u = 3 \times u_{opt} = 0.7$ mm/s (15 μm particle).

Typical maximum preparative injection volume is aproximately 50% of the peak volume. k' has been taken as 5 in all cases and $N = 2000$ plates/column.

$$V_r, \text{peak volume} = \frac{\text{flow rate (ml/min)} \times 1.7 \times (w_{1/2})\ (\text{mm})}{\text{chart speed (mm/min)}}$$

$w_{1/2} = t_r \left(\dfrac{5.54}{N}\right)^{1/2}$ from the plate number equation.

From $k' = \left(\dfrac{t_r - t_o}{t_o}\right)$

$t_r = t_o(1 + k')$

$V_r = \text{flow rate} \times 1.7 \times \left(\dfrac{5.54}{N}\right)^{1/2} \times t_o(1 + k')$

secondary. In the case of a strongly retained material and a weakly retained one the former will compete much more effectively than the latter for the sites of interaction at the top of the column and the latter will be pushed down the column and so

Ch. 1] Comparison between analytical and preparative liquid chromatography

effectively see a reduced column length. The solutes thus effect each others k' value, the weakly retained solute being affected more than the stronger. This is indicated in Fig. 1.1. When α is small, solute interaction is much more significant.

Required throughput
From a knowledge of the annual requirement of the solute of interest a throughput rate per month, per week, per day or per hour can be calculated. Dependent on the production method, batch or continuous, small scale or large, various options of column size, particle size, run time, etc., can be chosen. Table 1.13 gives an example.

Table 1.13 — Throughput

Consider the following simple example of a continuous production run.
  Annual output of compound of interest = 100 kg
  Weekly output = 2.22 kg (45 operational weeks p.a.)
  Daily output = 0.45 kg (5 days per week)
  Hourly output = 56.2 g (8 hours per day)

With a half-hour chromatographic run time this throughput could be accommodated on a column 600 mm × 100 mm i.d. The column contains approximately 3 kg of packing material.

With a batch production process the batch may be split and run as parts, or a large column may be deemed prefereable.

Mobile phase
The choice of mobile phase is of great importance in LC but in the preparative case additional factors need to be borne in mind. In total it is responsible for the following.

1 — Determination of the separation characteristics, degree of retention and chromatographic performance.
2 — It influences the operating pressure of the system, via its viscosity and diffusivity coefficient (see Table 1.14).
3 — It influences the cost of the purification process. Solvent costs can account for over 50% of the total purification cost of the chromatographic method.
4 — Because of the high volume of solvent used, it must be pure and free from non-volatile additives/impurities to prevent contamination of the compound of interest. Potential sources of solvent contamination, e.g. plasticizers from plastic tubing, shouuld also be avoided in the equipment.

Table 1.14 — Variation in diffusivity coefficient (D_m) with mobile phase

Mobile phase	Viscosity (cP, mNs m)	Temperature (°C)	Solute	Diffusity, D_m (m/s × 10)
Hexane	0.33	20	Toluene	3.8
Acetonitrile	0.35	20	Toluene	2.4
Methanol	0.60	20	Toluene	1.7
Water	1.0	20	Toluene	0.9
Methanol/water (40/60, v/v)	1.5	20	Phenol	0.74
Methanol/water (70/30, v/v)	1.48	20	Phenol	0.70
Methanol	0.40	50	Toluene	2.8

Thus, intrinsically low viscosity solvents or solvents at elevated temperatures increase the value of D_m which increases the optimum flow rate for the separation. This will have the effect of increasing the efficiency of the column when run at the high flow rate typically employed in preparative chromatography. This in turn will lead to higher throughput — the object of the exercise.

5 — Ease of solvent re-use/recycling should be borne in mind and the ease of mixing solvents to attain the initial composition is important. The use of single solvents is the ideal.
6 — Because of the volume of solvent used, toxicity and flammability properties should be studied (see Table 1.15).
7 — In most cases solvent will be re-used/recycled. If not, the cost of waste disposal should be considered.

The choice of solvent is therefore very important and may even determine which column to use; a case of choosing a suitable mobile phase and then finding a column to do the separation. In most cases isocratic runs will be performed because of the potentially increased throughput but with perhaps a solvent step change to remove late peaks. The option of gradient elution though is important.

Equipment constraints
This is often centred around the maximum flow rate and pressure rating of the equipment.

Specialized application constraints
This pertains to the individual application and the way it affects the purification step, e.g. batch size and frequency of manufacture, stability of the material etc.

Comparison between analytical and preparative liquid chromatography

Table 1.15 — Preparative LC mobile-phase solvents

Solvent	Flash point (°C)	Toxicity TLV p.p.m.	Boiling point (°C)	Comments
n-Hexane	−23	100	69	
n-Heptane	−4	400	98	
Methanol	10	200	65	
Ethanol	12	1000	79	Excise problems
Propan-2-ol	12	400	82	Viscous
Acetonitrile	6	40	80	Toxic
Ethyl acetate	−4.4	400	77	
Dichloromethane	—	100	40	
Dichloroethane	13	10	84	Toxic
Chloroform	—	10	61	Toxic
1,1,1-Trichloroethane	—	350	74	
Trichloroethylene	—	100	87	
Acetone	−18	1000	56	Not UV

TLV = Threshold Limiting Value

Physical properties of packing materials
— Composition of the material — silica, styrene-DVB, acrylamide, etc.
— Strength of the material — the particle should be robust and not degrade with use. Since preparative instrumentation may be run close to its specification limit any generation of fine material will yield unacceptable back pressure and force a reduction in flow rate thus extending run time and reducing throughput.
— Pore diameter — the pore diameter should be sufficiently large to allow complete diffusion of solute molecules into the pores of the material but not too high that the surface area is low. It is important that the pore size distribution is narrow since the small pores will be inaccessible to the larger solute molecules and so loading levels will be lowered. Too high a pore size is wasteful on space within the particle since it reduces the surface area of the support. Many small pores will degrade performance by reducing the rate of mass transfer between solute and the stationary and mobile phases.
— Surface area — the very large surface area of the supports is due to the pore structure of the particles. Basically the larger the surface area, the greater the load capacity of the particle. The surface area must be useable to be effective, the

presence of very small pores adds greatly to the apparent surface area but because they are too small for solute molecule penetration they are of no benefit.
— Pore volume — the pore volume affects the porosity of the particle and the pressure drop in the column and the rigidity of the particle. Typically the greater the pore volume the higher the porosity but the less rigid the material. It is important, however, to compare like with like since, during the manufacturing process enhanced strength and rigidity can be built into higher pore volume materials. A high pore volume is particularly important in size-exclusion chromatography. The pore size (PD), pore volume (PV) and surface area (SA) are related by the well-known equation.

$$PD(\text{nm}) = \frac{K \times PV(\text{cm}^3/\text{g})}{SA(\text{m}^2/\text{g})}, \quad \text{where } K \text{ is a constant.}$$

— pH of the silica — pH variations can confer different chromatographic properties on the silica and also different bonding characteristics on the materials.
— Presence of surface ions — the presence of various contaminants, e.g. iron, calcium, etc., can affect chromatographic performance and to some extent explain differences observed with silicas from different manufacturers.

Chemical properties of packing materials
The type of bonded phase used in the preparative run will match that used analytically although it may be different to that used in the initial analytical separation. Consideration should be given to developing a separation with optimum selectivity and resolution in the shortest time with cheap low-viscosity solvent. The bonded phase should be stable under the operating conditions employed for the same reason. It is important that the stability of the phase is known, especially with regard to pH range and temperature range, and that this stability is reproducible from batch to batch.

Simple strategy
— Optimize the analytical HPLC separation bearing in mind the additional constraints of the preparative situation.
— Maximize α to obtain the desired resolution.
— Determine the required value of column efficiency for the preparative peak of interest.
— Ascertain the required annual throughput and translate this to a required hourly or daily throughput.
— Translate this value into a choice of column sizes, material, particle size and flow rate.
— Determine the maximum load level of sample that may be injected from sample solubility data in terms of weight per volume injected.
— Cost the prep LC separation procedure.
— Scale up to the desired prep size, probably via an intermediate stage (e.g. 25 cm × 22.5 mm i.d.) refining as necessary.

— Test the prep system and evaluate yield, purity, throughput, etc.
— Obtain a final costing for the system.

It is almost certain that some fine tuning will be required and consideration should be given to the following areas:

— Feasibility of elevated temperature for the mobile phase to reduce viscosity. This will reduce back pressure and enable a higher flow rate to be used for no loss, and a possible gain, in efficiency by improving diffusivity and mass-transfer rates.
— Use of a lower viscosity solvent to give the same benefits as a temperature increase.
— Solubility of the sample components in the concentration in which they are loaded in the mobile phase. Much broader peaks than expected will occur if the solute comes out of solution at the head of the column.
— The diffusion properties of compounds will vary and so values of N should be based on the actual peaks being separated rather than ideal solutes. In all cases N will be lower than the ideal case, especially when biological molecules are involved.
— To prevent detector overload during scale-up trials if using an analytical detector either fit a prep LC cell or detune the operating wavelength.

For this simple strategy we may consider three preparative LC approaches:

Touching bands — the high yield and purity makes this approach ideal for R & D or small prep applications.
Overlapping bands— more applicable to larger samples.
Displacement — large process-scale application, specialized approach.

Table 1.16 compares these approaches which are illustrated in Figs 1.2 and 1.3.

It is not the object of this chapter to delve into the theory of preparative LC except to comment that an understanding of the theory will be of great benefit in preventing wasteful trial and error. The paper by Knox and Pyper [3] is a standard text and the other references will be very useful [2, 4–10].

Consideration should also be given to purpose-designed equipment and to alternative techniques. As mentioned earlier in 'System Operation', there are several additional techniques which, although available to the analytical chromatographer are not often used. They are however very useful in preparative LC to speed-up or optimize the run. Three of these will be considered in detail to illustrate the concept [8].

Backflushing
Backflushing is a technique used to remove slow-running solutes from a pre-column or guard column as an alternative to either gradient elution or column washing with a strong eluent. Both the latter techniques require the column to be re-equilibrated at the end of the run and in the preparative situation this increases cycle time and

Table 1.16 — Comparison of preparative HPLC approaches

Procedure	Features
Touching bands	— 99% + recovery of 99% + pure product — shorter run times — no reprocessing of impure fractions — minimal product loss during processing — smaller sample size, lowest production rate; well-suited for up to a few grams of purified product — no collection or analysis of multiple fractions; easy assignment of optimum cut points — easy development of optimum conditions for separations
Overlapping bands	— <99% recovery of pure product from each run — reprocessing of impure fractions from prior runs — larger sample size, higher production rate; well-suited for grams to kilograms of purified product — fractions must be collected and analysed; adequately pure fractions are pooled for recovery of product — optimum separation conditions require additional developmental effort
Displacement	— <99% recovery of pure product from each run — much longer run times (usually); reprocessing of impure fractions from prior runs; column must be regenerated after each run — much larger sample size, highest production rate; probably suited to kilogram-scale purification — fractions must be collected and analysed; adequately pure fractions are pooled for recovery of product — optimum separation conditions require a very considerable developmental effort; principles of separation not as well understood as for other procedures.

reduces throughput. The use of isocratic elution is both simpler and, with backflushing, potentially much faster. The detail of the technique depends on the actual chromatogram but broadly speaking backflushing removes slow-moving solutes from the head of the front column by pumping mobile phase the reverse way through the column.

Backflush can be performed using any column system but the outlet of the first column must use tubing that can withstand the back pressure generated by the

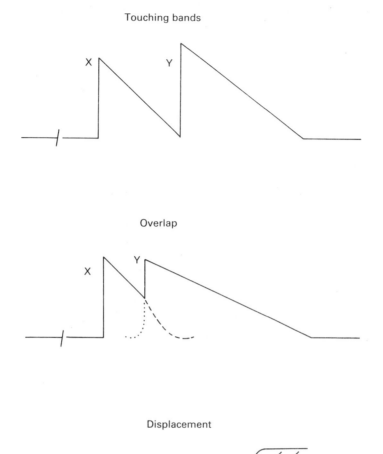

Fig. 1.2 — Preparative LC approaches. Touching bands, overlapping bands and displacement (Ref. [3]).

second column. Fig. 1.4 indicates the arrangement of the columns, 10-port valve, pumps and detectors. An example using a CEDI auto-axial compression cartridge system is shown with two pumps and either two detectors or one detector with two

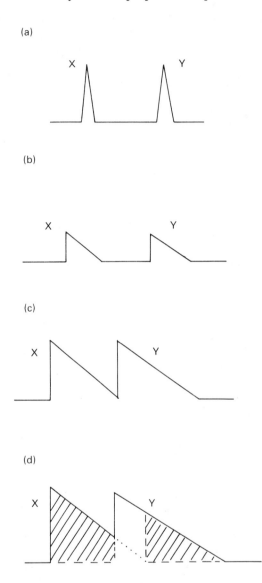

Fig. 1.3 — Idealized separation of two bands, X and Y as a function of sample weight: (a) 1 μg per band, (b) 2.5 mg per band, (c) 10 mg per band and (d) 20 mg per band (Ref. [4]).

cells. There are three distinct stages. The sample in the sample loop (or loaded via a through-pump valve) is injected into the first (guard) column. Flow is through the first column then the detector, the 10-port valve again, main column, second detector (or detector cell) and finally to the fraction collector. The backflush pump (P2) is off.

This status continues until the last peak of interest has passed though the first detector; i.e. fully eluted from the guard column and entered the main column. The

Ch. 1] Comparison between analytical and preparative liquid chromatography

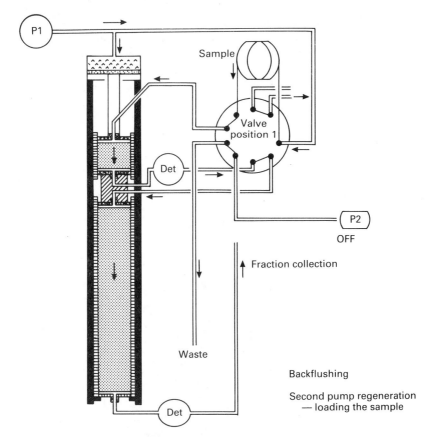

Fig. 1.4 — Backflushing: column and valve arrangement. Valve in position 1 — loading the sample onto the guard column.

10-port valve is switched to position 2, usually pneumatically or electronically. The peaks of interest in the main column continue to be eluted by the main pump (P1). The regeneration pump is switched on and backflushes the guard column with the same mobile phase as for elution and for the same time (plus a few seconds) as the initial elution of the guard column. This will ensure that all slow-running compounds still on the guard column are washed to waste, or possibly a collection vessel. The flow rate of P2 will typically be the same as for P1. If it is not the same, the backflush time must be adjusted so that at least the same volume of liquid is passed in backflush as in the initial elution. If loop sample injection is being used it can be reloaded at this time.

If it is desired only to collect the early peaks then it is possible to perform backflush with only one pump. The initial clean solvent, i.e. up to the t_0 position can be used to backflush the guard column. After this backflush step the valve is switched again to continue the elution. This is a very 'solvent and pump efficient' way of operating but is only applicable in some instances. It can be easily checked from the elution volumes and the scout analytical chromatogram.

34 Comparison between analytical and preparative liquid chromatography [Ch. 1

Sample pre-concentration and elution
In this technique samples are concentrated on a pre-column, then eluted onto the main column. Whilst elution is occurring the next sample is being loaded from the second pump onto the pre-column. In this way throughput is increased since no time is wasted loading a large volume of sample onto the column. Fig. 1.5 gives the system

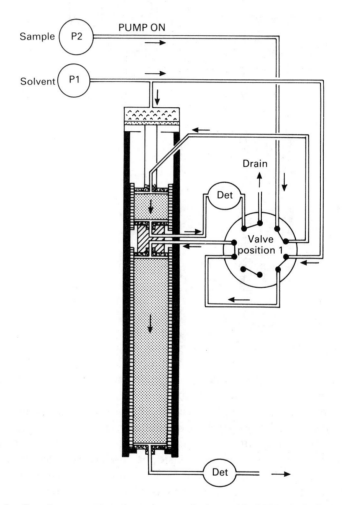

Fig. 1.5 — Sample pre-concentration: column and valve position. Valve in position 1 — loading the sample and elution in main column.

arrangement. Again, two site detection is used. The sample is loaded from a second pump. In Fig. 1.7, the sample is loaded onto the pre-column either to a preset amount or until breakthrough of the peak of interest occurs. The valve is switched to position 2 and the contents of the guard column are eluted onto the main column. The mobile phase supplied by P1 will always be stronger than the loading solvent. The peaks are detected as they leave the pre-column and enter the main column and are detected again after chromatographic separation on the main column. When all peaks have been eluted from the pre-column the valve is switched back to position 1. Elution in the main column continues but the next sample can now be loaded onto the pre-column ready for elution when the earlier run is complete. This system can be easily automated to chromatograph a large volume of sample.

When using sample pre-concentration and simultaneous elution the size of the pump should be matched in order to make the best use of time. The next sample should be fully loaded and ready to go before the elution of the previous sample has been completed.

Recycle

In this technique we use cartridges or columns in series and transfer the sample from the bottom of one cartridge to the top of the other in turn, i.e. 1 to 2, 2 to 1, 1 to 2, etc. In this way the effect of a long column with an increased number of theoretical plates can be mimicked. We are looking for an increase in the total number of plates for the column system.

Fig. 1.6 shows the arrangement of the system. After loading the sample in the loop it is injected onto column 1 by switching the valve to position 1. If second pump injection, or through-pump valve injection is being used, the sample would be loaded in this position. Chromatography occurs in column 1 and peaks are observed as they leave column 1 and enter column 2. When they have entered the second column the valve may be switched so that after leaving column 2 they re-enter column 1. Obviously we would not wish our pair of difficult-to-resolve peaks to mix with slow- or fast-running unwanted peaks so some pre-clean-up, e.g. blackflushing to remove slow-running peaks, or initial voiding of fast runners on the recycle system, may be needed. The valves are switched until the front of the faster-running peak on column 1 starts to run into the tail of the slower-running peak on the other column. In practice this is not allowed to happen since the front of peak 1 and tail of peak 2 are collected to minimize the recycled peak volume. In the example given in Fig. 1.7, collection would be from recycle 5 onwards.

Not all peaks that co-elute are suitable candidates for recycle. Under the conditions employed there must be a difference in their k' values. As usual it is best to determine suitability at the analytical scale prior to the preparative scale. Under the preparative conditions used the number of plates for the 20 cm cartridge can be determined. It will always be less than the optimum for the cartridge since it will be run typically in an overload situation.

Suppose for a 20 cm cartridge with 15 µm ODS we have, under the preparative conditions employed, 2000 plates at our disposal and our two peaks of interest run as one peak. A 25 cm × 4.6 mm i.d. analytical column with an equivalent 5 µm material may well, for the two components of interest, generate, say 64 000 plates/m or 16 000 plates/column, i.e. 8 times the number of the preparative cartridge with 15 µm material or 7 recycles.

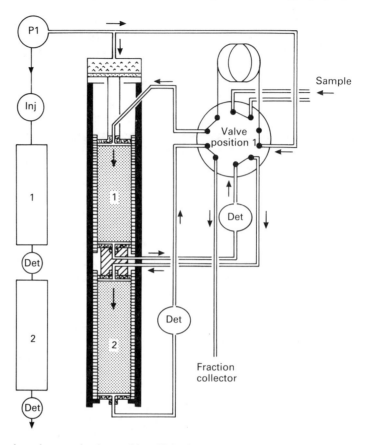

Fig. 1.6 — Recycle: column and valve position. Valve in position 1 — inject sample onto cartridge 1 and admit to cartridge 2. *Note*, position 2 switches stream from cartridge 2 to cartridge 1.

If the two components are fully resolved then they are candidates for prep recycle. If not then a change in chromatographic conditions to change selectivity is a better option. It should be remembered that resolution is a function of $\sqrt{N}$, so doubling the plates only gives 1.4 times the resolution.

Other valve techniques/combination techniques

— Sample concentration and backflush of pre-column.
— Two column, sequence reversal with column 1 backflushed.
— Front, heart and end cuts to second column or detector.
— Simple multi-column operation.

These techniques, which again use a 10-port valve, can be very useful for giving initial crude cuts as well as final runs.

Ch. 1]   Comparison between analytical and preparative liquid chromatography   37

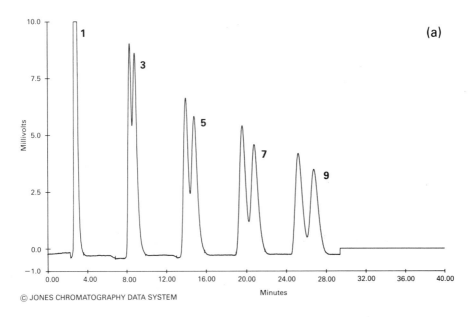

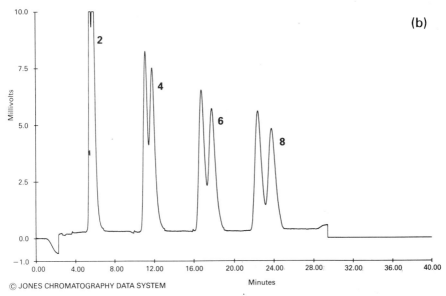

Fig. 1.7 — Recycle: 2 × 200 mm × 50 mm i.d. CEDI cartridges. Apex Prepsil 15 μm ODS; mobile phase 95% methanol/water; sample dimethylphthalate and diethylphthalate. (a) Cartridge 1. (b) Cartridge 2.

Adsorption/desorption techniques

This simple technique is similar to a solid-phase extraction technique and may be used to increase the yield of impure mixture and remove dissimilar compounds from the mixture. A simple low-pressure system may be used, for example: load sample onto column in weak solvent. Wash with weak solvent to remove one set of impurities. Elute product with intermediate-strength solvent. Other impurities are left on the column which is regenerated by passing a strong solvent. The partially purified product may now be further purified on a higher-performance column. The technique may also be used after the main run to concentrate the product in a more volatile solvent.

Frontal elution
This is a displacement technique whereby a strongly retained component pushes a weakly retained component down the column. It allows a much higher loading level of the column to take place. It will not be considered in detail here since it is covered fully in a later chapter.

Chromatographic technique combinations
If preparative LC is being used it may well be necessary to consider the use of more than one chromatographic technique. For instance, flash chromatography followed by high-resolution preparative LC has been used to isolate low-level synthetic bulk drug impurities [5].

In the separation of biological materials size-exclusion chromatography (SEC) is often used as an initial step in the isolation of high molecular weight species. This may be followed by ion-exchange and/or reversed phase with perhaps affinity chromatography, which has the best bioselectivity, as the final step.

In conclusion, in preparative LC one should try to look laterally at the chromatography problem in order to solve the real problem which is to maximize the purity and yield of the desired product in the minimum time and for the minimum cost with the maximum safety and convenience. The product protocol should be automatic in the true routine production of process situation and be within the capability of the production personnel and the resources of the project. The job is the isolation of the product, chromatography is merely the way of executing the job.

REFERENCES
[1] H. Colin, *International symposium on preparative and upscale chromatography*, Baden-Baden, Germany 1988.
[2] R. M. Nicoud and H. Colin, *LC-GC Int*. Vol. **3**, No. 2, 28 (1990).
[3] J. H. Knox and H. M. Pyper, *J. Chromatography*, **363**, 1 (1986).
[4] G. B. Cox and L. R. Snyder, *LC-GC Int*. Vol. **1**, No. 6, 36 (1988).
[5] R. M. Ladd and A. Taylor, *LC-GC Int*. Vol. **2**, No. 9, 53 (1989).
[6] P. E. Antle, G. B. Cox, S. I. Sivakoff and A. P. Goldberg, *Biochromatogr.* **2**, 46 (1987).
[7] G. Cretier, and J. L. Rocca, *Chromatographia*, **21**, 143 (1986).
[8] N. R. Herbert, lecture transcript *Preparative and Process Scale Liquid Chromatography* course, Loughborough University, England (Dec. 1989).
[9] F. Eisenbeiss, S. Ehlerding, A. Wehrli and J. F. K. Huber, *Chromatographia*, **20**, 657 (1985).
[10] M. Verzele and C. Dewaele, *Preparative High Performance Liquid Chromatography: a Practical Guideline*, Drukkerij de Muyter, Belgium (1986).

2

Chromatography systems: design and control

A. F. Mann
Amicon Ltd, Stonehouse, UK

INTRODUCTION

A chromatography system is fundamentally the same whether it is for analytical, small-scale preparative, or process use. The basic components (see Fig. 2.1) are common and consist of:

— the stationary phase or matrix
— the column to contain the matrix
— a pump to push mobile phase through the column
— means for mixing different solvents to produce gradients, either step or linear
— sample injection
— detection on the column outlet
— fraction collection
— control/data collection.

The particular use for a chromatography system will influence the relative importance and requirements of the individual components.

For instance, analytical systems where the objective is to identify components in the sample, have to be able to accommodate highly efficient columns containing very small particle diameter packings with small sample loadings. Here minimal volume in pipework, valves and detector cells are required and a large emphasis is placed on data handling. There is no requirement for fraction collection.

In contrast, preparative systems, where the objective is to obtain purified components of the sample, absolutely require fraction collection. However, preparative systems still vary in their requirements depending on their use. In process development where the system may be used in the investigation and development of many different purification problems, flexibility is paramount with the ability to operate with different media, solvents, columns and detection requirements.

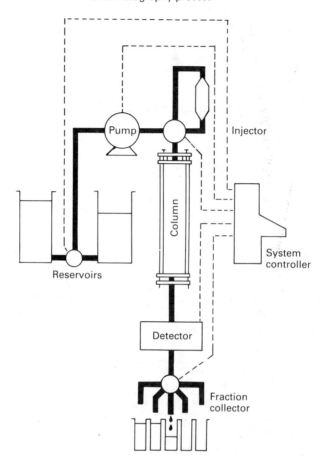

Fig. 2.1 — The basic components of a chromatography system.

In production where a system is dedicated to a single use, flexibility is no longer required and reliability will be the major concern.

SYSTEM REQUIREMENTS

To assist in designing a system it is useful to consider it having to meet a number of requirements as follows:

— function
— material compatibility
— pressure
— electrical

— hygiene
— control
— reliability
— serviceability

Function

The chromatography separation is based not on the system but on the sample interaction with the stationary and mobile phases. However, if the potential of the separation is to be maximized the system must not introduce deleterious effects into the process. Consequently functional requirements involve not only selecting the right number of solvent inlets, fraction outlets, pump and valve types, sensors detectors, etc., it also involves ensuring that pipework configurations and sizes are optimal for the required flow and that dead legs and dilution zones are kept to a minimum.

Material compatibility

It is important to ensure that no problems can arise from adsorption to, or leaching from, components within the system. Materials need to be compatible with all solutions used in the process, including regeneration and cleaning. Particular attention must be paid to the use of plastics or elastomeric seals with organic solvents. The effect of solvents is dependent on concentration, temperature and contact time.

Attack usually results in softening and dissolving of the plastic. Consequently the mechanical loading on the component also needs to be considered. Ideally, tests should be performed with samples of the actual components proposed, to be certain of acceptability. For these reasons HPLC systems invariably use only stainless steel and PTFE as construction materials.

In the case of aqueous systems where halide ions are present, stainless steel of at least 316L grade is required, and in the case of high concentrations of chloride even more resistant grades may be specified.

It is also necessary to ensure that materials used comply with any regulatory requirements (e.g. those of the Food and Drug Administration (FDA)) where appropriate.

Pressure

Here it is necessary to ensure that not only are all components within the system appropriately rated but that the design of the system adequately covers the pressure drop in the column and the associated pipework and valving for the flow required. The major contribution to pressure drop in the system is invariably the packed matrix bed. Information on this can be derived from the early process development trials and from the matrix manufacturer.

Caution should be exercised to ensure that data relates to performance within the column diameter proposed. With rigid particle packings, data from small-diameter columns (25–50 mm diameter) can be readily extrapolated to larger diameters. In the case of soft or deformable gels this may not be the case owing to the fact that in the small-diameter column the bed is supported by the column wall, the effect of which is lost in larger diameter (>200 mm diameter) columns. This results in larger than anticipated pressure drop in the larger columns. In some cases the matrix may be so deformable that planned flow rates cannot be achieved in large-diameter columns, owing to the compressibility of the matrix.

It is important to remember that the pressure drop is related to flow rate. This can be particularly important when designing a system with a wide flow-rate range. In order to minimize dilution within the system at low flow rate it is desirable to use small-bore tubing and valves. This may, however, produce an unacceptably high pressure drop at the high flow rates, necessitating a compromise between minimum system volume and pressure drop.

In the case of the packed bed the pressure/flow relationship is linear, whereas with pipework it is a square factor where a doubling in flow rate will result in a fourfold increase in pressure drop.

The implications of pressure vessel regulations also need to be considered, especially with respect to the column. It may be necessary to ensure these are constructed and certified to specific standards (e.g. ASME).

Electrical safety requirements

Besides the need to meet appropriate standards for dust and water protection (IP or Nema rating), the use of organic solvents can mean that the equipment needs to be designed and built to comply with regulatory standards for explosion proofing, enabling it to operate in a hazardous area. Sometimes the chromatography itself is not using flammable solvents but it is installed in a hazardous area where flammable solvents are being used for other processes. In this case the chromatograph will still need to be built to comply with explosion-proofing regulations.

Although in the past each country had its own standards and regulations, for instance BASEEFA in the UK and PTB in Germany, there are now unified European standards under the auspices of CENELEC (European Committee for Electrotechnical Standardization).

Summary of IP protection numbers

First number — Protection against solid objects
IP Tests
0   No protection.
1   Protected against objects up to 50 mm, e.g. accidental touch by hands.
2   Protected against solid objects up to 12 mm, e.g. fingers.
3   Protected against solid objects over 2.5 mm (tools and wires).
4   Protected against solid objects over 1 mm (tools, wire, and small wires).
5   Protected against dust-limited ingress (no harmful deposit).
6   Totally protected against dust.

Second number — Protection against liquids
IP Tests
0   No protection.

1. Protection against vertically falling drops of water, e.g. condensation.
2. Protection against direct sprays of water up to 15° from the vertical.
3. Protected against direct sprays of water up to 60° from the vertical.
4. Protection against water sprayed from all directions — limited ingress permitted.
5. Protected against low-pressure jets of water from all directions — limited ingress permitted.
6. Protected against low jets of water, e.g. for use in shipdecks — limited ingress permitted.
7. Protected against the effect of immersion between 15 cm and 1 m.
8. Protects against long periods of immersion under pressure.

Explosion-proofing designations

Explosion-proof equipment designated Ex or EEx (the latter signifies compliance with Cenelec Standards), is classified under three categories, namely:

Class: Zone 0 1 2	This is the classification of the area in which the equipment can be used. Most equipment is rated to Zone 1 which is an area where an explosive gas–air mixture is likely to occur in normal operation. Zone 0 is where an explosive gas–air mixture is continuously present and Zone 2 is an area where an explosive mixture is unlikely to occur, and if it does so it will only exist for a short time.
Group: 11A 11B 11C	Classification to the explosion characteristics of the materials for which the apparatus is suitable. Most chromatography solvents will fall within the 11A category.
Temperature rating: T1–T6	This is the maximum surface temperature that will be experienced. T1 <450°C T4 <135°C T2 <300°C T5 <100°C T3 <200°C T6 < 85°C

In addition the rating plate will indicate the type of protection used.

      d — flameproof enclosure
      p — pressurized enclosure
      i — intrinsic safety
      e — increased safety.

Hygiene

The purification of sample feedstocks derived from micro-organisms or natural products, coupled with the use of mobile phases that are designed to maintain biological activity, often present ideal conditions for contamination of the system and proliferation by unwanted micro-organisms and the generation of pyrogens.

Effective cleaning, and if necessary sterilization, is required in these circumstances if a product of the required level of purity is to be obtained.

The need for sanitary designed systems is primarily restricted to low- or medium-pressure systems. The solvents, matrices and sample feedstocks encountered in HPLC are less conducive to bacterial growth.

Process hygiene is not only important in terms of preventing contamination of the required product but also in prolonging the life of the stationary phase. The most commonly accepted protocols for sterilizing chromatography matrices are by cleaning in place (CIP), with strong alkaline solutions.

For CIP procedures to be effective there must not be unswept areas within the wetted confines of the system. In such systems for instance 'Tri Clamp' style pipework fittings will replace threaded or ferrule-type connectors, as the face-to-face seal with flush-fitting gasket does not provide crevices for bacterial growth. In contrast, ferrule fittings invariably provide a 'dead area' between the tube and outer fitting in front of the ferrule. Similarly in the case of valves, diaphragm types are preferred as the design permits free flow across the whole internal surface in contrast to ball valves where not only can there be a contaminating 'plug' within the ball itself, but the area between the ball and packing seal may form a crevice for bacterial growth.

In some situations, however, it is not possible to select required components that meet the above criteria. However, it may still be possible to meet required hygienic conditions with the introduction of specific additional cleaning procedures that may require isolating, stripping down and cleaning separately those components that are not satisfactory cleaned by standard CIP procedures.

In the design of sanitary systems attention has increasingly been focused not only on the need for effectively flushed fittings and connections, but also on the surface of the pipework, valve or column itself, in terms of being non-conducive to bacterial or fungal attachment and growth.

Increasingly therefore the surface of stainless steel pipework and fittings in contact with the process stream are being specified with an electropolished finish. The surface of stainless steel, even if highly polished mechanically, is in fact not smooth but consists of a series of peaks and troughs. Electropolishing has the effect of removing the peaks, thus not only reducing the total surface area available for bacterial attachment but also reducing the crevices, formed by the peaks being bent over during mechanical polishing, which can trap bacteria or other material.

Control

Here it is necessary to consider and decide the level of control and automation required. The main reasons for automating processes is to reduce cost, by decreasing operator involvement and hence labour cost, increasing throughout by for instance operating twenty-four hours a day, which can reduce capital investment and ensure reproducibility and reliability. In addition, GMP and Regulatory compliance may impose documentation requirements that can more easily be met by automated systems.

The degree of automation, and in particular the user interface, will depend on the process for which the system is required. A system used in development or in production, but for different applications, will need more flexibility than that used in dedicated single-step production process. Here reliability and simplicity in the user interface is more important.

Reliability
Reliability is absolutely paramount in a production system, in contrast to a development or research unit where this may be sacrificed for capability or flexibility. The location of the system, for instance a cold room, together with the stability and quality of services (power, air) need to be considered. Particularly with automated systems in industrial situations, the susceptibility to and implications of power drop outs and other potential electrical problems need to be considered.

Serviceability
Systems should be designed with the aim of minimum service requirements but routine adjustments and calibration of sensors will be necessary. In production it is preferable to implement routine preventative maintenance to reduce the likelihood of a component failure during a production run. The requirements and frequency of routine servicing should be advised by the manufacturer or compiled from the individual component manuals.

SYSTEM DESIGN

Having considered the overall system requirements, the next step in system design is to produce a functional or block flow diagram of the process.

The process flow and functional specification requires an understanding of not only the chromatography but also the upstream and downstream requirements of the process. This stage will define column sizes, together with solvent and fraction tank volumes and CIP requirements.

This then leads to the establishment of a 'P&ID' or piping and instrumentation drawing, that defines all pipework, valving, instruments and process functions, together with the battery limits of the system. The P&ID then forms the basis for selecting the necessary components or equipment needed to perform the required functions.

COMPONENT SELECTION

Column

The objective of the column is to not only contain the matrix but also allow it to be packed homogeneously and provide uniform flow throughout the whole cross-sectional area of the bed, in order to ensure maximum capacity. At the same time as little dilution as possible is desirable, either within the bed itself, or more importantly at the inlet and outlet.

In analytical columns a single inlet and outlet is made to the column with the packed bed contained between two mesh or sintered frits. When the sample is

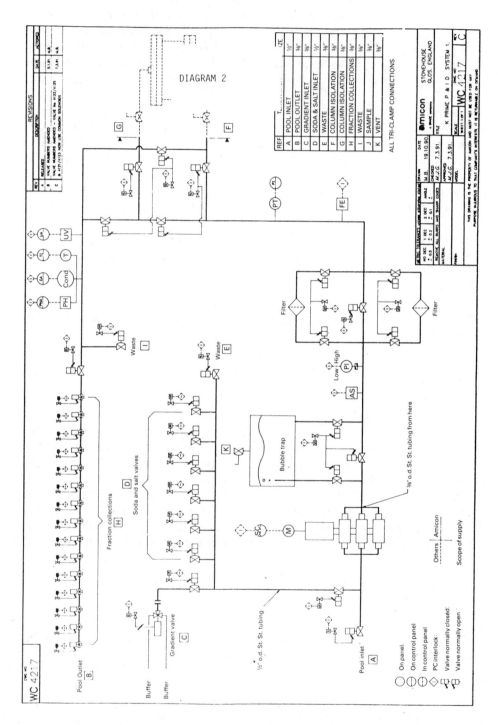

Fig. 2.2 — Block flow diagram of a chromatography system (P&ID).

injected onto the column through single inlet port, some radical dispersion may occur in the frit, but often the column behaves in the infinite-diameter mode. This is where the sample enters as a single-point injection, which as it flows through the packed bed, disperses radially but may not contact the wall before exiting through the bottom frit. This can be beneficial in analytical columns as the failure to contact the walls negates any peal broadening due to the different flow profile existing between the bed and the column wall, than within the bed. This results in sharper peaks.

From a preparative perspective this type of flow profile is not so desirable as it reduces the effective capacity of the column. In this case it is more important to ensure use of the complete bed so it is necessary to ensure uniform presentation of the sample across the whole cross-sectional area of the column. Although this will result in some peak broadening, due to the wall effect, this is reduced as the column diameter increases.

Several designs have been used to assist radial distribution. The simplest of these consists of a single inlet/outlet port with a coarse mesh interspersed between the end cell and a finer mesh retaining the bed. The coarse mesh acts to provide channels for radial distribution. On larger diameter columns this approach is augmented by having several inlet/outlet ports manifold together.

An alternative approach has been to use a machined end cell containing distribution channels radiating out from a single inlet-outlet port. A bed support consisting of a mesh or sinter material is placed between the end cell and the bed. A further refinement, whereby the distribution channels are machined in the form of a cone with the apex at the inlet/outlet port, helps to prevent the possibility of air locks forming, which could give rise to non-uniform flow.

Selection of the column depends on the type of matrix used and the pressure and solvent requirements. For low-pressure chromatography (1–5 bar) there are commercially available columns, constructed of plastic, glass or stainless steel, ranging in size from laboratory scale up to 2 m diameter (glass maximum 1 m). Glass and transparent plastic columns permit visual inspection of the packed bed and may be considered advantageous both for packing and running. Glass columns may, however, be seen as not being robust enough for industrial use, while plastic columns may not have the required solvent compatibility. Stainless steel, although robust and solvent resistant, is unsuitable for products that are susceptible to metal ions. Also, high concentrations of halide ions can attack stainless steel.

In general, low-pressure columns are packed by pouring the matrix gel slurry into the column and either letting it settle under gravity, or by flowing mobile phase through the column to settle the bed. This may require the use of a column extension tube or filler tube during packing, which is removed once the bed is settled. The use of a column with an adjustable end cell greatly facilitates this operation, as it means that accurate measurement of the gel volume is unnecessary, as is the case with a fixed-volume column. In addition if any further settling occurs during running, then adjustment can be made to take up any void.

Process HPLC columns are invariably made of stainless steel because of the high pressures involved. In addition the more difficult packing of the smaller HPLC matrices, compared to larger particle materials, results in these columns usually being associated with ancillary packing/unpacking equipment. The most widely used technique is that of axial compression, although alternative approaches such as radial compression have been used.

In axial compression the slurry is introduced into the column and then compressed, expelling the excess solvent. In small-diameter columns the bed may be stable for many runs but in larger diameter columns (>75 mm), particularly with spherical matrices, the bed may settle further, giving rise to voids and loss of efficiency. In these cases the use of equipment offering constant bed compression will alleviate this problem, as the end cell will automatically adjust to take up any voids.

Columns, in particular HPLC ones, may be classified as pressure vessels and be subject to regulations which differ from country to country. Pressure rating *per se* is usually not an indication as to whether the column is regarded as a pressure vessel, but rather the product of pressure and volume. The larger the column, therefore, the lower the pressure above which it would be classified.

Pumps

There are a number of pump types from which to select, including:

— peristaltic
— centrifugal
— gear
— lobe
— plunger
— diaphragm

and the following factors need to be considered.

Flow rate/pressure

Most pumps will provide a 10:1 turndown ratio in flow through speed control of the pump motor. This is usually sufficient on dedicated production units, but in the use of diaphragm or piston pumps where stroke length can be varied, together with the use of single or multiple pump heads, can considerably increase the range.

Flow rate must be stable and reliable over the entire range and outlet pressure must be sufficient. For this reason positive displacement pumps are to be preferred, particularly at high pressure. However, on large-flow-rate systems (10–20 l/min) centrifugal, gear or lobe pumps offer significant cost advantages over diaphragm pumps. Under such circumstances the non-linear flow characteristics can be overcome by having automatic control of the pump via a flow meter feedback.

One disadvantage of positive displacement pumps is pulsation of the pump output. It is undesirable for this to reach the column, as it can cause disturbance and loss of efficiency of the packed bed. For low-pressure systems appropriately sized bubble traps, placed on the outlet of the pump can act as pulsation dampeners. High-pressure systems require a specific pulsation dampener.

Chemical/temperature compatibility
All materials in the pump, as elsewhere in the system, must be compatible with all process solutions, including all cleaning chemicals. If steam sterilization or hot CIP cycles are used then the pump must be appropriately specified, to ensure seizing does not occur owing to component expansion.

Hygiene
Where sanitary aspects are critical the pump head should be easily dissembled for cleaning.

Sample compatibility
Most proteins can be damaged by shear, and although the chromatography process usually employs a single pass, in contrast to ultrafiltration, it may nevertheless be desirable to select low-shear pumps. Lobe and peristaltic pumps have lower shear characteristics when compared to centrifugal or gear pumps.

Reliability
Reliability of pump seals is crucial to ensure either no loss of product or no contamination of the process stream. Peristaltic pumps, whole offering a 'very clean pump', have the disadvantage that the tube may split. This risk can be reduced by routinely replacing the tube.

On high-pressure diaphragm pumps the diaphragm is usually driven by a hydraulic fluid. Possible contamination of the process stream, owing to rupture of this diaphragm, can be removed by specifying double-diaphragm pumps. In this case two diaphragms are fitted together with a sensing element that detects if either of the diaphragms has ruptured. This will enable an alarm to be signalled whilst also allowing the completion of that process run.

Valves
Where possible, multipoint valves are preferred as they minimize dead volumes. Ball valves are well suited to this application, especially the use of three-way valves for fraction collection and four-way valves for filter and column by-pass. However, sanitary aspects may dictate that only diaphragm valves are used. In this case two-way valves will need to be manifolded to provide the same function. Careful attention must be paid, however, to minimizing dead volume and orientating the valves to ensure free draining. High-pressure systems invariably use ball valves as diaphragm valves are limited in pressure capability.

Gradient
The mixing of two solvent-inlet streams, either to produce step-wise concentration changes or linear changes, may be required by the chromatographic process. Alternatively the ability to dilute concentrated buffer solutions by mixing with water can save on tank and space requirements.

Gradient or solvent mixing is usually performed in one of two ways. Either by switching a valve between the two solvent streams on the inlet side of the pump, the duration of the valve position being in relation to the ratio of solvents required, or by using individual pumps for each solvent operating at different flow rates and blending the outputs of both pumps together. This latter approach can be used with piston or diaphragm pumps where automatic stroke length adjusters are used to vary the ratio of flow output from each pump head. The pump heads can then be driven from a single motor shaft, the speed of which can be varied to alter total flow output.

Bubble traps/air sensors
Bubble traps, besides acting as pulsation dampeners, also serve an important role in protecting the column from air inclusion. Bubble traps are only really necessary on low-pressure systems with the 'soft gel' matrices. Small amounts of air introduced into a silica-based matrix will have no deleterious effect on the matrix, although it may result in flow irregularities. In contrast the 'soft gel' matrices will dry out causing cracking in the bed and serious flow irregularities. Even if the air is subsequently pumped out of the column the matrix will not revert properly and the column must be unpacked, the matrix properly reslurried and wetted and then repacked.

The bubble trap removes small amounts of air that may be introduced during connection and disconnection of buffer lines, etc. As an added precaution it is usual to fit an air sensor after the bubble trap which will alarm and shut down the system should the bubble trap fail to trap air or become empty of liquid.

Air sensors can also be used on the sample-tank outlet to ensure complete emptying. Air sensors have the advantage over tank-level sensors in that they do not need to contact the sample and can ensure complete emptying of the tank as they are on the outlet pipe rather than in the tank itself.

Several types of air sensors are available, based on optical, capacitance or ultrasonic designs. Only optical or ultrasonic designs can be used on distilled water as capacitance systems require a minimum conductivity of the fluid.

Filters
All solutions being passed through the system must be free of particulate matter. Its accumulation leads to clogging of the column bed or bed support, resulting in high back pressure and non-uniform flow. Filters can either be incorporated in the chromatography system or solutions prefiltered. If filters are incorporated, it is usual to employ separate filters for the sample and mobile phases. Filtration should ideally be down to 0.22 μm to help prevent bacterial contamination.

Liquid filters may also act as air traps as, once wetted, the bubble point of the filter may be sufficient to prevent air being passed. If filters are incorporated, maximum economy of filter replacement requires that differential pressure monitoring be installed to indicate plugging of the filter.

Pipework
The pipework must meet the requirements for material compatibility, pressure rating and be designed to give minimal pressure drop, while at the same time giving minimum internal volume. Sanitary requirements will dictate internal finish and type of connections.

Instrumentation

Parameters most commonly monitored include flow, pressure, UV, pH and conductivity. In addition temperature may also be monitored and refactive index detectors may be used in processes where non-UV-visible adsorbing compounds are being purified.

Flow

Flow meters are used not only to monitor flow but also in conjunction with a controller to regulate pump speed. For sanitary systems, Magflow meters are preferred but these will not operate with deionized water or solvents, and are not available for small pipework systems ($<\frac{1}{4}''$). In these cases turbine meters are usually used.

Pressure

Pressure sensors are used to monitor not only the pressure of the system but also to alarm or initiate control, if for instance the pressure rises due to filter or column plugging, or the pressure drops due to a failed connection.

UV/visible adsorption monitors

Most biologically active compounds, whether proteins or syntehtic drugs, adsorb light in the UV or visible spectrum and so can be detected by adsorption monitors. In development a variable-wavelength detector is desirable for its flexibility, but in production a single wavelength monitor is preferred, from both cost and reliability standpoints. Most systems will use 254 nm for detection of small molecular weight synthetic components and 208 nm for proteins.

In preparative systems, selection of the correct pathlength can be important to prevent saturation of the detector. Detectors with variable pathlength cells are, therefore, useful for development. Identification of peaks on the UV monitor output can be used by the controller to initiate fraction collection.

pH/conductivity

pH and conductivity detectors are used not only for monitoring conditions during the chromatography run, but also for automating regeneration, cleaning and equilibration steps. pH electrodes are particularly sensitive to fouling. Although double-reference-electrode self-checking systems are available they are not very compact and result in large internal-volume flow cells. Single-electrode systems are, therefore, most commonly used, but these must have provision for easy removal of the electrode for cleaning, calibration and replacement.

Conductivity sensors are either of the contacting type or the induction type, which works on the principle of the solution coupling the magnetic field of two magnetically isolated induction coils and has the advantage of not having electrodes, therefore, there is no polarization and no decrease in performance owing to formation of deposits. The disadvantages are that these units are currently available in sizes that require relatively large hold-up volume fittings.

In selecting sensors particular attention must be paid to any explosion-proofing requirements, and to the ease of sensor calibration by the operator.

AUTOMATION

The simplest chromatography system can consist of manually operated valves, a variable-speed pump, detector and chart recorder to log the detector output. As already discussed, however, significant advantages can be obtained by introducing varying levels of automation. This might range from just gradient generation and fraction collection, all the way through to full process control of sample injection column regeneration, CIP cycles, etc., together with automatic alarm handling and documentation of process variables.

The end use of the chromatographic system is important in determining the level and type of automation required. Process development requires a flexible automation system, easily easily configurable to different hardware use and enabling rapid method generation and modification. In contrast, a dedicated production unit requires simplicity in user interface, ideally just a start/stop switch and reliability.

Automation can be approached in essentially one of three different ways, or combinations of these; namely, dedicated controller, programmable logic controller (PLC) or computer-based systems, all of which are based on a microprocessor chip.

Dedicated controller

Here a controller is designed and built for a specific application. This approach is usually used where there is a market or requirement for several identical controllers. In this case the cost of designing an setting up the manufacture of a specific controller is offset by lower unit cost through multiple manufacture. The dedicated controller has the advantage that it is usually easy to operate and ready for use. A disadvantage is that flexibility for enhancement or modification is invariably limited.

A dedictated-controller approach can either be applied to the complete chromatographic system or to sections of it, for instance a gradient controller, which in turn can be integrated into the overall control via a personal computer.

Programmable logic controller (PLC)

PLCs have long been used for industrial control purposes, being considered rugged and reliable for these environments. PLCs are usually available in modular form to enable units to be connected together to provide control of the required number of input/outputs.

PLCs require specific extensive programming by skilled personnel, either through a PC or a dedicated interface. For this reason PLCs tend to be used in such a way that the basic programme does not require changing once commissioned. System parameters that need to be adjusted during the process or from process to process, are dealt with by separate dedicated controllers that can easily be adjusted by the operator. PLCs are, therefore, more suited to production applications then development situations.

Computer-based systems

Here a computer, usually PC-based, in conjunction with an I/O interface is used to control and monitor the system. The I/O interface may be directly linked with the computer or be provided in the form of a dedicated controller or PLC. The advantages of the computer-based system depend very much on how the additional power of this approach is used.

More extensive data logging can be used, recording not only system variables, but also events such as valve switching and alarm conditions. Similarly, software packages can be used that will enable the system to perform all the functions of a PLC, but with the ability to allow rewriting of method files with no programming skills required.

The various functions required of a controller can be considered as follows:

— manual override of all output functions
— automatic switching of valves/events on the basis of time, accumulated flow or dependent on sensor outputs
— gradient generation
— control of — flow
            — pressure
— process status display — valve position
                          — sensor readings
— process data logging of all — variables
                           — events
— actions off variables
— peak detection
— alarm display and reports — pressure
                          — flow
                          — pH
                          — conductivity
                          — air
                          — valve position
                          — tank level
— feedback control of valve positioning
— safe power-up–shutdown sequencing
— simple programming and editing
— security of access to different levels of control limited to authorized personnel
— output — printer/plotter
          — floppy disk
          — network
          — data transfer to other software programs for report writing
— environment — IP rating
                — explosion proofing

The relative importance of the above requirements will differ in respect to end use of the system. In production more emphasis will be put on reliability by ensuring simplicity of operator interface, increase in alarms and security of access to various levels of control.

VALIDATION

GMP and regulatory compliance require process validation. The FDA defines process validation as follows:

> 'Process validation is establishing documented evidence which provides a high degree of assurance that a specific process will consistently produce meeting its pre-determined specifications and quality characteristics'.

It is not sufficient, therefore, to merely ensure that the raw materials of a process, or the materials of construction of process equipment, meet required specifications of purity and traceability; it is also essential to test and document the function of all equipment including the control system.

Validation, in effect, begins right back at the design stage as it is at this point that the specifications and operational limits of the system must be established.

This preliminary step ensures that the hardware will be suitable for its intended function, and will not be adversely affected by its operating environment. Procedures for system testing must be established and documented as part of the validation. The test methods used must be designed to ensure the system meets specifications and will operate reliability. Test procedures must test the equipment at the limits under which it will be required to operate. This approach is known as 'worst-case testing'. This does not imply testing to the point of system failure, it is testing only to the outer ranges of the operational limits that are established by the user.

If the components have been chosen correctly in respect of the user's requirements, worst-case testing should not unduly stress the equipment.

Information obtained from these studies should be used to establish written procedures covering equipment calibration, maintenance, monitoring and control.

CONCLUSION

In conclusion, the design of a chromatogaphy system requires careful consideration as to end use of the equipment in terms of capability, flexibility and reliability. Once these requirements have been decided, individual components or systems can be selected bearing in mind at all stages the need, in a production situation, for process validation.

3

Technical structure of liquid chromatography separation plants: on the pilot and production scales

G. Munch
Darmstadt, Germany

INTRODUCTION

High-performance liquid chromatography (HPLC) has been used analytically for many years as a high-resolution separating procedure. It provides both qualitative and quantitative information about the composition of the substances being examined.

In recent years ever more stringent standards have been applied to the purity of many chemical and pharmaceutical products, and this has resulted in increased use of preparative liquid chromatography for purification purposes. Improvements in chromatography systems, with regard to both procedures and apparatus, have resulted in more efficient separation, selectivity and throughput. It is now possible to obtain high purities and greater yields. The different methods of chromatography offer users a wide field of application for preparative liquid chromatography (PLC) through the use of various physico-chemical properties as separation criteria.

As will be shown in the following, the requirements of the different methods of chromatography must be taken into account in the design of plants for preparative liquid chromatography. The process flow diagram, the type of protection of the electrical equipment, and the materials used are chiefly affected. Furthermore, the discontinuous liquid chromatography process must be automated by arranging individual separations in series, so that a semi-continuous process is created. Only in this way can purification be performed economically.

THE DEMANDS MADE ON PREPARATIVE CHROMATOGRAPHY PLANTS

The design of preparative liquid chromatography plants is directly derived from that of analytical plaants. There are many different types of preparative plants, incorporating widely varying equipment.

The technical demands on a chromatography plant depend on the particular separation problem, and thus in large part on major influential factors such as raw product, separation method, packing material and mobile solvent. (Fig. 3.1).

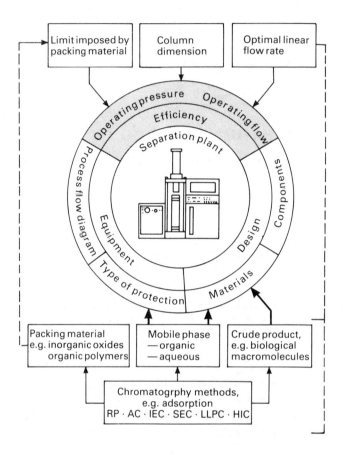

Fig. 3.1 — Demands made on a preparative liquid chromatography plant.

If the raw product contains, for example, bio-polymers, there is a high degree of risk that they will be denatured by metal ions or organic solvents. In such a case special demands are made on the materials that come into contact with the product. The material used must be absolutely stable (e.g. resistant to organic solvents or salt solutions), therefore the choice of material is also influenced by the mobile solvents used. If large amounts of organic solvents are employed, the plant must be explosion-proof.

The operational pressure and the throughput of the plant are largely influenced by the optimal linear rate of flow and the dimensions of the columns used. Because the packing material is only limitedly resistant to pressure (e.g. in the case of polymer gels <10 bar), flow and thus separation can be impaired.

Technical structure of liquid chromatography separation plants

The major factors influencing the design of preparative plants are therefore the raw product and the mobile solvent (Fig. 3.2). Plants are then differentiated according to the materials that come into contact with the product (stainless steel and inert plastics) and the type of protection (explosion proof and non-explosion proof).

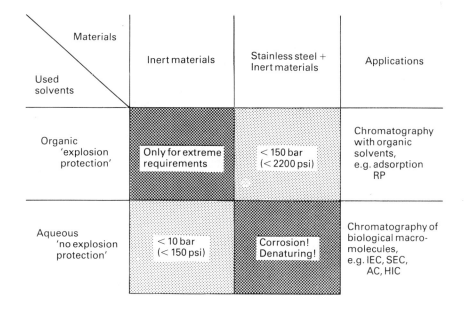

Fig. 3.2 — Types of plant design for preparative liquid chromatography.

If both criteria are combined, there are four possible types of plant design. Depending on the solvents employed, two areas of application can be distinguished:

— chromatography of low molecular compounds with organic solvents;
— bio-chromatography, more precisely chromatography of biopolymers with aqueous buffered solvents.

In the former case, plants are generally made from stainless steel and are explosion proof. Relatively small molecules are usually separated, and the danger of structural changes being brought about by metal ions is slight. The use of non-metallic materials in association with organic solvents is therefore only appropriate for extremely demanding applications.

Chromatography with aqueous solvents requires no precautions to be taken against explosions. The main priority here is inert materials (materials that do not interact in any way with the substances to be separated). One of the results of structural changes caused by metal ions is that the molecules suffer a loss of biological activity. In addition, corrosion problems can be expected if metallic materials are used. For these reasons, highly resistant fluoroplastics or glass are usually employed as the materials that come into contact with the product.

Stainless steel plants are usually used for adsorption or reversed-phase (RP) chromatography, where extremely pressure-resistant organic carriers are employed. The molecules to be separated are relatively small and therefore possess large coefficients of diffusion, so that high linear flow rates are possible. If fine grain materials are used there is a high pressure drop across the column. Stainless steel plants are therefore designed to withstand high pressures up to 150 bar.

By contrast, bio-polymers have small coefficients of diffusion. Linear flow rates are therefore low and a much lower pressure drop can be expected. As already mentioned, the pressure resistance of the packing materials is often below 10 bar, so that inert plants are usually designed for this field of application.

PREPARATIVE PLANTS THAT OPERATE WITH ORGANIC SOLVENTS
Plant concepts

Fig. 3.3 shows the major components of a preparative production plant to be operated with organic solvents. The dimensions of the individual units are based on the need for high throughput and continuous, and therefore economic operation.

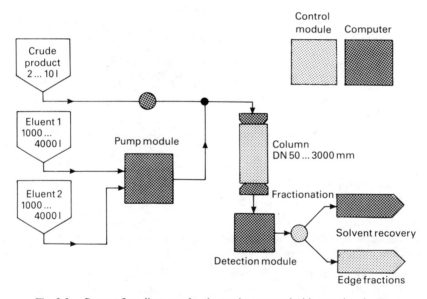

Fig. 3.3 — Process flow diagram of a plant to be operated with organic solvents.

One or more eluent tanks are needed to accommodate the daily throughput of 4000 l or more, along with an eluent pump, whose capacity must be suitable for handling these amounts.

In contrast to analytical applications, where amounts of raw product measured in ml and nl are fed in, feed volumes can comprise several litres. For this reason the receiver must also have an appropriate size.

The preparative columns through which these amounts of liquid flow, and in which pressures of up to 150 bar can be generated, especially with fine materials,

Technical structure of liquid chromatography separation plants

have a correspondingly large diameter of up to 300 mm and more. The detection unit must also be capable of handling the higher throughput in preparative applications, where eluate streams of 6 l/min and more must be detected.

In view of the large volumes of solvent involved, the control and automation unit must be installed in an explosion-proof area. The principal tasks of the control system will be dealt with in the section, on Plant Controls.

Without now going into the structural differences between the individual components, it is possible to construct preparative plants for a variety of concepts from the named parts. The more important will be dealt with in more detail later.

One of the alternatives is to feed the raw product directly through the suction side of the eluent pump into the column. The suction-side valves must be regulated to deliver eluent and raw product to the head of the column in differently timed cycles.

The advantage of this concept is the relatively simple design of the plant, the possibility of operating with a low-pressure gradient, and the possibility of varying the feed volumes of raw product. This constellation is suitable for applications where feed volumes of raw product are large.

In the case of separation problems where the feed volumes of eluent and raw product are in the ratio 1000:1 it is virtually impossible to meter the very small amounts of raw product through the relatively large eluent pump with any degree of accuracy. Because there is an unavoidably large dead volume in the solvent pump, there is a danger that the peak bands will be widened, with an associated loss of chromatographic definition.

To improve the relatively inexact metering of the raw product in the low-pressure area, the raw product can be fed through its own pump at the high-pressure side of the column head (Fig. 3.4). The pressure and flow ranges of the raw product pump must be adapted to the eluent pump.

If it is desired to improve the mixture ratios for gradient operation, a change is made from the illustrated low-pressure gradient to the high-pressure gradient shown in Fig. 3.4.

Because the mixture ratios in the low-pressure range of the eluent pump can be subject to considerable flow and pressure fluctuations, it is difficult to reproduce the flow conditions in the low-pressure range. This can be combatted by mixing the eluent stream on the high-pressure side with another pump head.

Components

The individual components of a plant can exhibit various designs and structures. For this reason the requirements of the specific application must be taken into account when components are selected.

The major demands made on an eluent pump are:

— very precise metering, even with fluctuating back pressures,
— exclusion of non-permissible over-pressures,
— pulsation-free eluent stream, to prevent stress on the column packing,
— explosion proof.

Displacement pumps are generally used for preparative chromatography with organic solvents on account of the high flow rates and pressures involved. Especially

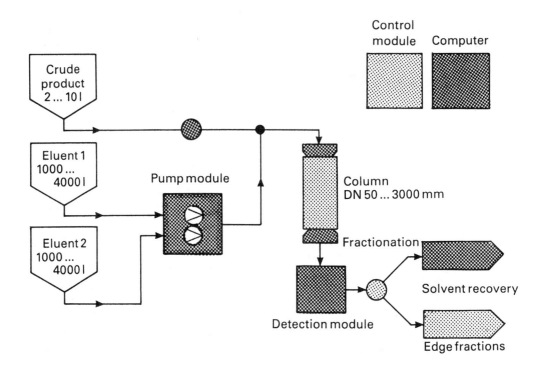

Fig. 3.4 — Process flow diagram of a plant with separate product feeding and high-pressure gradient pump for operation with organic solvents.

widely used are diaphragm pumps which in principle, as illustrated (Fig. 3.5), are fully leakproof compared with piston pumps. Because the transport medium and transport unit are kept completely separate from each other, seals cannot suffer wear and impurities cannot get into the eluent.

As already mentioned, the raw-product feed system must be able to deliver variable, but exactly reproducible, amounts of raw products to the head of the column. Flow and pressure conditions must be adjusted to the eluent flow. This is why displacement pumps must be used, because they are best able to comply with this requirement. Not only time-regulated but also flow-regulated diaphragm and piston pumps are employed.

Although diaphragm pumps possess the above-mentioned advantaage of being leakproof with regard to the carrier medium, the structure of piston pumps is such that they can deliver more-exact displacement volumes.

The valves used in fully automated plants can be categorized as suction side, high-pressure side and low-pressure side valves, depending on where they are located

Ch. 3] Technical structure of liquid chromatography separation plants 61

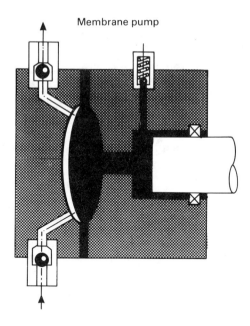

Fig. 3.5 — Essential design of a diaphragm-metering pump.

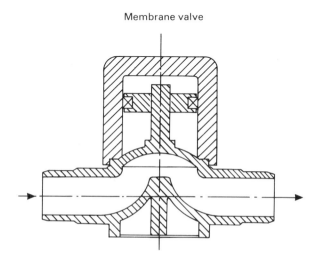

Fig. 3.6 — Essential design of a diaphragm valve.

(Fig. 3.6). Valves located before the pump, i.e. on the suction side, are characterized by large pipe cross-sections at relatively low pressures. Pneumatically regulated ballcocks and diaphragm valves are most commonly encountered. They permit short switching times and almost turbulence-free flow.

High-pressure side valves are located between pump and column. Only pneumatically regulated ballcocks are capable of maintaining the high degree of sealing efficiency required at the pipe cross-sections and high pressures associated with preparative chromatography.

Pneumatically regulated diaphragm valves or magnetic seating valves are especially suitable for the low-pressure region behind the column. The lack of dead volume makes them easy to clean.

PREPARATIVE PLANTS THAT OPERATE WITH AQUEOUS SOLVENTS

The dimensions of a chromatography plant for aqueous solvents are shown in Fig. 3.7. The pump usually has a capacity of up to 300 l/h, and the column has a diameter of up to 1 m. The receiver and eluent tanks have a capacity of approx. 4000 l to support continuous operation round the clock.

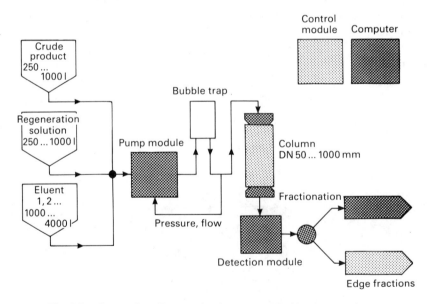

Fig. 3.7 — Process flow diagram of a plant operated with aqueous solvents.

The general concept associated with aqueous solvents differs in a few respects from that applied to organic solvents.

Low-pressure gradients are almost always employed in conjunction with static or dynamic mixers, because it has been found that the quality of the mixture so obtained is adequate. Generation of the gradient on the low-pressure side is economically more favourable, because only one single-head pump is needed. Furthermore, it is possible to generate gradients with more than two eluents at the pump without additional effort, by using several differently timed valves. Besides the eluent receivers, tanks for regenerating solutions are attached, so that, for example, sterilization stages with sodium hydroxide solution (0.1 — 1 molar) can be incorporated into the production run. This means that, in CIP (cleanning in place) procedures, all parts of the plant can be automatically rinsed between separation cycles.

The practice of suction side feeding of the raw product through the eluent pump has become established with aqueous solvents. This usually presents no problems because chromatographic separation methods such as ion-exchange chromatography (IEC), hydrophobic interaction chromatography (HIC) and affinity chromatography (AC) are concentration processes, i.e. large feed volumes of highly diluted raw product can be handled. The need for precise raw-product metering is not as great as in the case of small volumes of concentrated raw product, and feeding through the eluent pump is not critical.

The pump is regulated by flow or pressure. This means that on the one hand the flow can be stabilized, and on the other hand a low-pressure column (e.g. made of glass) can be operated at the limits of its pressure range, so that optimal throughput is achieved.

Depending on the flow rate, gas bubbles in eluents may collect in the head of the column and impair the distribution of the liquid, or they may be rinsed through the column and disturb the polymer gel packing. For this reason a gas bubble trap should be located between the pump and the column to ensure that the eluent that reaches the column is bubble-free.

Besides UV absorption, the pH and conductivity of the eluate are also measured when aqueous solvents are used. pH and conductivity measurements provide important information for assessing the state of the column, e.g. after regeneration, while UV signals are an indicator of the product concentrations in the eluate.

Components

The demands made on the eluent pump (Fig. 3.8) are largely the same as in the case of organic solvents. For this reason displacement pumps are generally used, as is the case with organic solvents, but flow pumps are also used (for example, or centrifugal pumps). Absolute separation of the media, and operational pressures up to 10 bar, even in the case of pumps made from inert materials, are the major factors in favour of employing diaphragm metering pumps. As already mentioned, the pump must have a flow-volume regulation. This can be achieved by altering the stroke or adjusting the rotational speed. The pump must be safeguarded against overpressure. Here, however, the critical factor is not the maximum operational pressure of the plant but the maximum possible pressure that can be applied to the column used (e.g. 2 bar for a DN 450 glass column). Pulsation-free flow is just as important as in plants that operate with organic solvents. A flow-through diaphragm pulsation damper can be employed to damp pulsations.

A bubble trap (Fig. 3.9) is a container with eluent input and output below and a vent valve at the top. Undissolved gas in the eluents collects at the top of the container. If the level of liquid in the bubble trap falls below a defined minimum

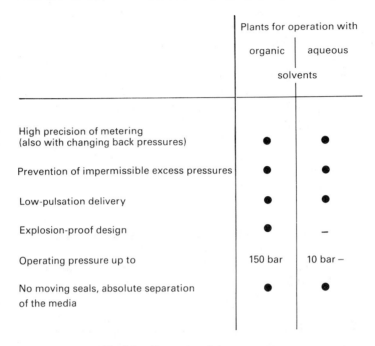

Fig. 3.8 — Properties of eluent pumps.

(filling level sensor), the bubble is automatically vented, so that continuous operation is possible. If the level of liquid in the bubble trap is regulated in such a way that a certain volume of air is always present, the compressible gas acts as a pulsation damper and contributes to more-uniform flow.

If possible, the valves should have no dead volume, in order to exclude product entrainment and to guarantee complete rinsing in CIP processes.

A distinction must be drawn between high-pressure and low-pressure valves. Diaphragm valves can be used at the low-pressure side. Maximum operational pressures at the high-pressure side can be as high as 10 bar, so that only ballcocks come into consideration for meeting the high demands with regard to being leakproof and easy to clean. The inevitable dead volumes associated with the exclusive use (e.g. 4-way valves for column reverse rinsing).

The transducers for pressure, flow, UV absorption detection signals, pH and conductivity must also be free of dead volume (i.e. flow through). Splitting the stream to measure UV, pH and conductivity should be avoided, because eluate entrainment could occur.

The flow meter should be as independent of viscosity as possible. The use of rotary vane transducers has proved popular in practice.

PLANT CONTROLS

The control unit of a preparative plant for organic or aqueous solvents must fulfil a number of requirements. A distinction can be drawn between general requirements and special chromatographic requirements.

Ch. 3] Technical structure of liquid chromatography separation plants 65

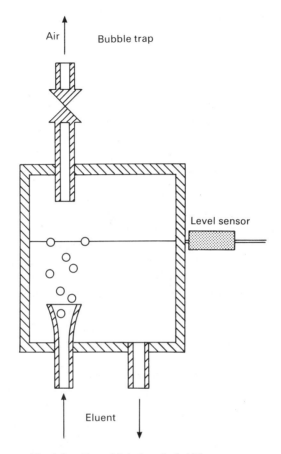

Fig: 3.9 — Essential design of a bubble trap.

The central task of the control system of a preparative plant is to ensure that the plant operates continuously round the clock by executing the discontinuous processes in sequence. To do this it must recognize when a separation is completed and then initiate the following separation cycle.

It should be possible to incorporate user-specific responses to changes in the separation process. Recognition of events such as product or disturbance peaks, base-line shifts or other plant disturbances should be error-tolerant. This means that the plant can also react in a controlled manner to events that lie outside the realms of reproducibility. After a separation programme error, for example, column rinsing could be initiated before final switch-off, to prevent loss of raw product.

Automatic recognition of the eluate peaks has a high priority in preparative plants. There are a number of different methods available:

— time-dependent fractionation
— volume-dependent fractionation
— detection signal-dependent fractionation, with controlled peak recognition.

Table 3.1 — Requirements for the control of preparative chromatography plants

General requirements

— Controlling and monitoring a reproducible process
— Documentation and administration of process data
— Visual representation of the process, with current process flow diagram and representation of various process parameters
— Operator-friendly interface with a variety of user levels
— Alarm concept for different plant disturbances
— If needed, explosion-proof link for working with organic solvents

Special chromatographic requirements

— Continuous sequence of separation cycles
— Automatic peak recognition (amplitude and minimum)
— Intelligent response to disturbances in separation sequences
— Real-time operation (rapid response to events)
— Capable of being integrated into higher-order systems (process control systems) for example, to ensure coordination with solvent processing

In practice, two types of peak detection predominate:
— flank detection (also known as level or time-amplitude detection) and
— minimums detection (or max-min detection)

Flank detection involves examining the detection signal for a certain amplitude value within a pre-specified time 'window'. This enables an ascending or descending flank of a peak to be fractionated.

If a mixture of substances is incompletely separated, a minimum search can be made. The curve is characterized by three points. The detection of individual substances can be made more reliable by logical combination of the events associated with various detection signals.

SUMMARY

Preparative liquid chromatography plants can be differentiated according to whether they operate with organic or aqueous solvents. The materials that come into contact with the raw materials play a major role. Plants for use with organic solvents are made of stainless steel and are made explosion proof, while plants for use with aqueous solvents are usually made from inert plastics.

The alternatives illustrated in the process flow diagrams require different forms of implementation. If only one product is involved, a clear plant concept can be drawn up, limited to the major requirements of this one product. A more complex but also more flexible concept allows a large number of individual separation problems to be solved. In any case the control concept must be flexible, to permit adaptation to changing process runs.

If all of the named demands made on a preparative plant are taken into account, production can be optimized, thereby enabling costs to be reduced by increasing throughput and yield.

REFERENCES

[1] H. Krauss, and M. Leser, Ingenieurtechnische Aspekte der Hochleistungsflüssigkeitschromatographie in der Produktion, *Swiss Chem.* **9**, 5 (1987).
[2] W. Reese, Optimale Steuerung für präparative HPLC-Anlagen, *Swiss Chem.* **9**, 6 (1987).
[3] F. Eisenbeiss, and K. Reichert, Grundzüge der präparativen LC, *GIT Supplement Chromatographie*, **3** (1987).

4

The design of preparative chromatographic media

Dr I. Chappell
Crosfield Chemicals, Warrington, UK

INTRODUCTION

Interest in preparative chromatography is growing rapidly as the long-awaited move to establish it as a major unit process operation gathers momentum [1]. This trend has been evident in the increase in symposia on the topic of preparative chromatography and in the change of emphasis fom theoretical to practical aspects of the technology [2].

In analytical liquid chromatography, the reversed-phase mode of operation accounts for the majority (60–70%) of applications. This preference for reversed-phase operation can be ascribed to the relative simplicity of the solvent systems used in the reversed-phase mode. Binary solvent mixtures of water and alcohol (or acetonitrile) are sufficient for the elution of most compounds and in addition re-equilibrium on changing solvent or on gradient elution is rapid. In contrast, on silica a wider selection of solvents are needed to cover elution range and careful conditioning of the silica column is needed for consistent chromatography.

In spite of the above, silica is still the most widely used media for preparative chromatography of low molecular weight (<2000 Daltons) organic compounds. This is particularly the case for large-scale processes using 10^3–10^4 kg p.a. The prime reasons for this preference are:

 (i) relatively low cost of silica vs reversed phase,
 (ii) good solubility of compounds in the typical solvents used on silica,
 (iii) easier removal of solvent from collected fractions and
 (iv) limited availability of reversed phase media at the 10^3–10^4 kg level.

With the above in mind, this chapter will look at the factors which have to be considered when designing silica media for preparative chromatography. Although brief mention will be made to reversed-phase media, the design aspects of the latter are beyond this short discourse.

Several factors need to be considered in designing a range of media specifically for preparative chromatography. The choice of manufacturing route and specification of parameters such as surface area, pore size, pore volume and particle shape contribute to the development of products with superior preparative performance across a broad range of applications from small organic molecules to biopolymers.

PRODUCTION ROUTES TO PREPARATIVE SILICAS

Consideration of end-user needs such as:

— reproducibility
— loading capacity
— flexibility in physical parameters

leads to the conclusion that large scale (1000 kg) batch production of irregular media using a dedicated facility is the preferred route to high-performance preparative media.

That large batch sizes are required for process-scale use is relatively obvious, but the choice of batch production may be less so. Conventional large-scale silica production is a continuous process with the gel setting time being governed by a trade-off between throughout (short gel time) and hardness (long gel time). Diverting silica gels from this type of plant for use in chromatography tends to give soft silicas whose specifications and quality are governed by the demands of its technical grade usage. For chromatography use, a higher specification and more robust silica can be manufactured by using a dedicated batch process with the specification and gel setting time governed by the requirements of the chromatography market.

The production of irregular vs spherical particles was also considered at an early state. The irregular route is a scaled-down version of conventional multi thousand tonne silica production. Silica gel in the mm size range is produced, milled and screened to the required particle sizes which have identical physical and chromatographic properties, the only difference is particle size. Thus a family of products ranging from 500 µm to 5 µm can be readily made and the yields of each fraction controlled to match the market demand.

Spherical silica production (excluding conventional spray drying which tends to give weak products) is essentially a small-scale operation. Until recently, 5 kg was a reasonable batch size of 5 to 10 µm media. Scaling up is intrinsically more difficult than scaling down. Spherical production gives a distribution of spheres which need classifying to give the required product. The dilemma on scale-up is whether to aim for a wide distribution from which a range of products can be classified or a narrow distribution which gives a higher yield of one size fraction. The problem with the former is that the yield of the various fractions may not match the market requirements. With the latter, different process conditions need to be developed for each product. These problems are additional to those normally faced on scale-up. What works well in a 5 l reactor making 5 µm spheres does not necessarily work on the 5000 l scale with a 30 µm sphere, the result is more likely to be a rice grain shape rather than a sphere.

Overall, spherical particles are more suited for analytical use (5 kg batch, 5 µm high sales value) than for preparative use (1000 kg batch, >10 µm, low sales value).

DESIGN FACTORS FOR PREPARATIVE MEDIA

A range of factors need to be optimized in the development of media specifically targeted at preparative chromatography, the major ones being particle size and distribution, pore volume, surface area, pore size and particle shape.

Before discussing the influence of each of these factors, a few comments on measurement techniques are worthwhile. Various techniques are available for measuring the properties of porous media. Many of these suffer both from limited relevance and limited accessibility to the end user. However, two measurement techniques stand out as being both relevant and readily available.

Column permeability measurement can be calculated directly [3] from chromatography data (Eq. (4.1)) using Darcy's Law [4] and equated to an equivalent particle diameter (Eq. (4.2)). The latter, together with column efficiency, allows the chromatographer to judge whether the performance of a particular material matches the quoted particle size.

$$\text{Specific permeability, } K^o = \frac{\eta L u}{P} \quad 10^{-9} \text{cm}^2 \tag{4.1}$$

η = solvent viscosity (cP)
L = column length (cm)
u = solvent liner velocity (mm/s)
P = pressure drop (bar).

$$\text{Equivalent particle diameter, } d_p = (1000 K^o)^{1/2} \text{ (microns)} \tag{4.2}$$

Values for a range of materials calculated from different manufacturers' published data are given in Table 4.1.

In general there is good agreement between the permeability-derived particle size and the claimed particle size. Any considerable discrepancy should lead to an investigation for blocked frits or excess fines.

Pore size and distribution can be measured by a variety of techniques. For pores up to 300 Å, nitrogen adsorption/desorption can be used, while for larger pores,

Table 4.1

Phase	$K^o \times 10^{-9}\,cm^2$	d_p (microns)
Sorbsil C60 5	0.25	5
Sorbsil C60 10P	1.2	11
Sorbsil C60 15/20	2	17
Sorbsil C60 20/40	5	25
Sorbsil C60 40/60	40	60
Spherical 100 Å 12.5 µm	0.25	5
Ireregular 90 Å 10 µm	1.2	11
Spherical 80 Å 10 µm	1.0	10
Irregular 60 Å 10 µm	0.71	8.5

mercury porosimetry is more appropriate. Although suitable for the study of the physical structure of the gels, neither of these techniques can be considered to effectively mimic the permeation of large organic molecules into the pore structure which is the parameter of most interest in selecting media of the most appropriate pore size. However, the techniques of size-exclusion chromatography is an excellent model; it provides data in a form readily appreciated by the chromatographer [5]; it is easily carried out using standard chromatographic apparatus; it is applicable for all pore sizes and can also be used for many bonded phases (Fig. 4.1).

Particle size (µm)

The earliest preparative chromatography was carried out using relatively large particle sizes in the average particle size (aps) range of 50 to 100 µm, but in recent years there has been an increasing tendency to use smaller particles in the range 10 to 30 µm. The debate on the optimum particle size for preparative chromatography is undecided with theoretical studies indicating that throughout is directly proportional, independent or inversely proportional to particle size depending upon the factors incorporated into the model [6–8].

However, some evidence in selecting particle size can be gained by considering the various factors which are governed by particle size.

— *Resolution* is inversely proportional to the square root of the particle size (Fig. 4.2). Improvements in resolution are best made by optimizing selectivity. However, for complex mixtures, a high peak capacity is required, as altering selectivity is likely to lead to overlap with other components in the sample.
— *Pressure drop* is inversely proportional to the square of particle diameter. The significant rise in operating pressure as particle size decreases leads to substantial increases in the capital cost of the preparative chromatograph.
— *Ease of packing* decreases along with decreasing particle size. Above 30 micron aps, dry packing can yield effective columns but below this; electrostatic

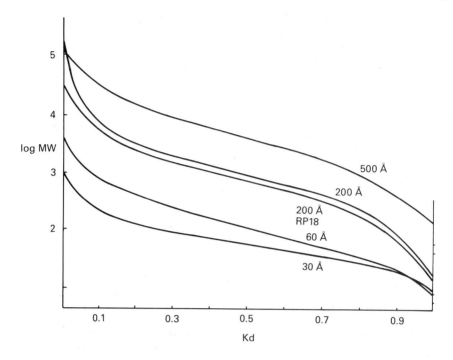

Fig. 4.1 — Size-exclusion chromatography of polystyrenes using THF indicates the molecular weight range which can penetrate the porous structure of the media.

attraction requires a modified procedure. For preparative columns, a number of approaches have been developed with varying degrees of success including modification of the slurry-packing procedures used to prepare analytical columns, radial column compression and axial column compression [9]–[11].
— *Cost* of packing increases sharply below 30 μm owing to the increasing difficulty in obtaining narrow particle size distributions. Typical costs are given in Table 4.2.

This demonstrates that from a media point of view, 20/40 μm silica gives the highest resolution per unit cost. For bonded phases, the cost of reproducible bonding reduces the price differential and leads to a more gradual increase in cost as the aps is reduced.
— *Loadability*. On the basis of percentage reduction of capacity factor or efficiency with sample mass, larger particles have a higher loadability (mg of sample/g packing) than small particles. Plot of efficiency vs. sample mass show that packings tend to a common efficiency as sample mass is increased.

However, from the viewpoint of throughput, it appears [12] that the degree of difficulty of the separation defines a minimum value for the term d_p^2/L (d_p=particle diameter and L=column length). Thus any combination of particle size and column length that exceeds this value can be used. Smaller particles will give quicker cycle times, less time on the column for sensitive samples, but more runs for a given throughput requiring more analysis and quality control testing.

Ch. 4] The design of preparative chromatographic media

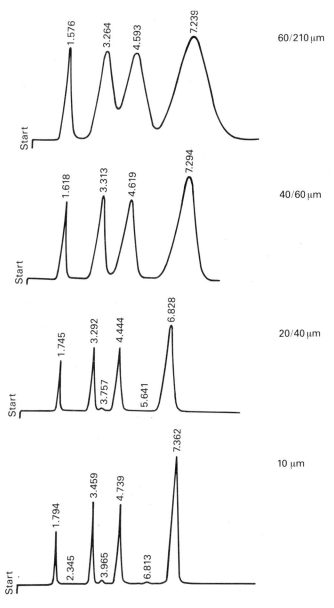

Fig. 4.2 — Reduction of particle size leading to increased resolution demonstrated on 60 Å silica (Sorbsil C60).

The debate on which particle size to use looks set to continue and as with many aspects of chromatrography, there may be no optimum, just a series of potential solutions which work well and which take into account the total cost of separation over the lifetime of the project.

Table 4.2

Particle size (μm)	10	15/20	20/40	40/60	60/120
Cost (£/kg)	600	200	33	33	25
Resolution ($\sqrt{N}$)	7.1	5.5	3.5	1.8	1
Resolution/cost ($\times 10^{-2}$)	0.1	2.8	10.6	5.5	4

Pore volume (ml g^{-1})

As a generalization, loading capacity of a phase is proportional to the available surface area per unit column volume. High pore volume gives substantially higher surface area per unit mass of packing, but this is counteracted by a lower packing density. The overall effect is to give higher surface area per unit column volume (Fig. 4.3) and higher loadability (Fig. 4.4).

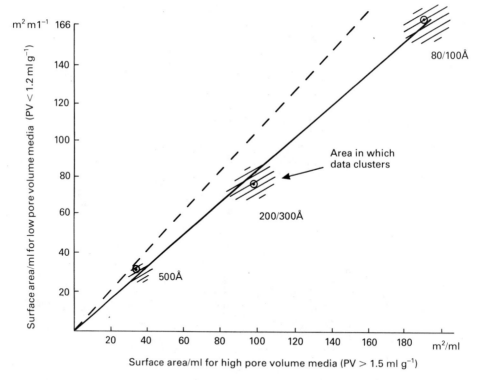

Fig. 4.3 — High pore volume media leads to higher surface area/ml of packed bed and higher loadability.

The lower packing density of high pore volume media gives considerable saving on media costs.

The one drawback of high pore volume media is its reduced mechanical strength. Although not a major problem with low-to-medium pore sizes (up to 200 Å), beyond this, stability at pressures greater than 4000 p.s.i. can be problematical for small

Ch. 4] The design of preparative chromatographic media

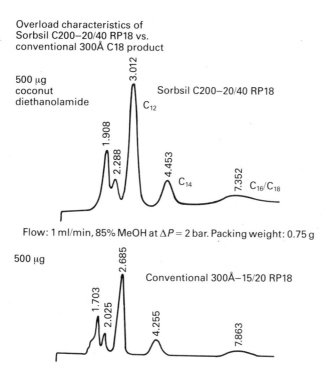

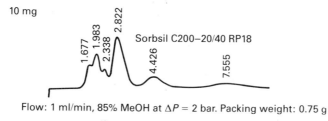

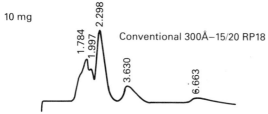

Fig. 4.4 — Higher phase loading of high pore volume media (18 %C) vs 8 %C) gives increased loading capacity.

particle media (e.g. 10 µm). Thus the use of small particles can preclude the loading advantages which high pore volume media can deliver.

Conversely large particle versions of media that were designed for stable operation as 3 and 5 µm packings will have low pore volume and lower loading capacity against a similar sized high pore volume media which is not constrained by having to be able to withstand a 10000 p.s.i. packing pressure.

High pore volume media with their high surface area per unit weight allow very high bonded-phase loadings. For example, a 200 Å high pore volume octadecyl phase will typically have a higher carbon content (18–20%) then a 100 Å low pore volume phase (typically 15% C). The high carbon loading of the reversed-phase media necessitates the use of higher levels of organic modifier in the mobile phase. This gives two added benefits:

(i) reduced solvent viscosity (in most cases) and
(ii) increased sample solubility in the mobile phase. This allows higher sample concentrations to be prepared.

Pore size (Å)

The choice of pore size for a particular separation is relatively straightforward in that the smallest pore size possible should be used. The size is determined by the size of the largest molecular to be separated. For each molecular size, pores smaller than a particular size will lead to a decrease in efficiency owing to slow pore diffusion kinetics, together with reduced loading capacity owing to exclusion from the smaller pores. Using media of a pore size larger than necessary leads to lower loading capacity owing to the lower surface area of wide pore media.

The correct choice of pore size can be illustrated by consideration of two common misconceptions.

Small molecular chromatography is often carried out on Å media in the belief that 60 Å media is the minimum limit for efficient chromatography. However, use of 30 Å media not only allows efficient separations for molecules of less than 600 Daltons, but the increased surface area and packing density can give up to a 50% increase in loading capacity per unit column volumn leading to higher throughput and higher concentration fractions (Figs 4.5, 4.6, and Table 4.3).

The result of this can be a throughput increase of 50% on a current plant with no capital investment and no increase in consumables. For a new process, considerable savings via smaller equipment sizing and reduced solvent costs can ensue. These savings can be upwards of 40% on consumable costs compared with conventional 60 Å media.

The equation relating to solvent savings alone are given by:

$$\text{Savings (£ p.a.)} = abce\,(1.38 - 1.14[1/(1+d)])$$

where a = mass of 60 Å silica used per annum,

The design of preparative chromatographic media

Aromatics on Sorbsil C30–10

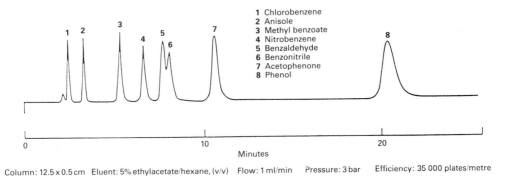

Fig. 4.5 — High-efficiency separations are achievable on 30 Å media for molecules of molecular weight less than 400 Daltons.

Loading capacity

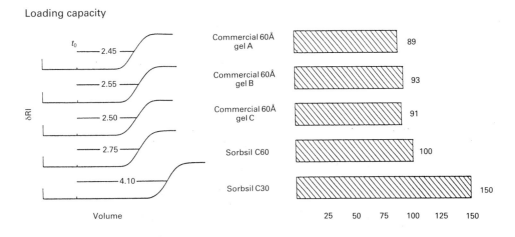

Fig. 4.6 — High surface area and packing density of 30 Å media can give substantial increases in loading capacity.

b = number of column volumes of solvent per run,
c = cost of solvent/l,
d = fractional increase in loading capacity.
  (Normally 0.5 for Sorbsil C30), and
e = number of times column is used before replacement.

Table 4.3

Media	Pore size (Å)	Surface area ($m^2 g^{-1}$)	Packing density ($g\,ml^{-1}$)	Surface area ($m^2 ml^{-1}$)
Spherical silica	120	220	0.62	130
Spherical silica	100	340	0.50	170
Irregular silica A	60	500	0.55	275
Irregular silica B	60	540	0.55	297
Sorbsil C60	60	550	0.55	303
Sorbsil C30	30	700	0.63	441

Likewise, it is commonly stated that 300 Å pore-diameter media is required for peptide and protein separations. It is readily demonstrated that high-capacity media of pore size 150 to 200 Å (Fig. 4.7).

A few observations can be made about pore size distribution. Much importance is often made of narrow pore size distributions. If all samples contained components of the same size, the use of a monodisperse pore size medium with a pore size of just slightly greater than the minimum required would be advantageous. In the real world though, samples contain a range of species of different sizes and relaxation in the pore size distribution can be used to good advantage in providing increased surface area for smaller components. How far the distribution can be relaxed before performance deteriorates will depend upon the particular sample to be separated.

Particle shape

The debate about the merits of spherical or irregular media has swung decidedly towards the former over the past decade as spherical media have become the choice for analytical separations.

In the analytical area this preference is well founded, particularly as the particle size has dropped to 3 μm, a size at which the higher permeability of spherical media is crucial in keeping the back pressure and consequent thermal gradients to a manageable level.

It has naturally been assumed that this preference for spherical media also holds for preparative chromatography. This assumption merits close scrutiny.

Given identical particle size distributions, the increased permeability of spherical media should result in a pressure drop of approximately a half that of irregular media. However, a comprehensive study of the efficiency of spherical vs. irregular media compared data on commercial columns packed with a wide range of spherical and irregular media produced over a period of years [13]. The large data base showed that while 5 μm spherical media had higher efficiencies than irregular media, the opposite was the case for 10 μm and above. Thus, in the particle range of interest to preparative chromatography i.e. >10 μm, a smaller particle spherical media is

Ch. 4] **The design of preparative chromatographic media** 79

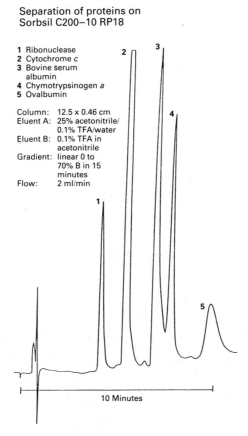

Fig. 4.7 — Peptides and proteins up to 100 kDaltons can be readily separated on 200 Å pore-size media.

needed to match the efficiency of an irregular media and the pressure-drop difference is substantially reduced.

Another feature of major importance in preparative chromatography is loadability, which is a function of surface area and directly relatable to throughput. Irregular silicas have surface areas two to five times higher than typical spherical silicas which makes the former particularly attractive for preparative chromatography.

Further points for consideration are:

— *surface chemistry*. Routes to irregular silicas lead to high surface levels which have been reported to the beneficial in producing stable reverse-phase media [14].
— *cost*. Irregular media cost significantly less.
— *availability*. Irregular media are available in 1000 kg quantities.

— *quality*. When made on dedicated plant irregular media are of high consistent quality.
— *ease of packing*. Currently 10 µm spherical media have a better reputation for ease of packing into large columns but this could reflect the extent of experience with these media. Irregular media appear to need different packing regimes and studies on these are less advanced than for spherical packings.
— *range of particle sizes*. Irregular media from 5 to 220 µm aps, spherical media seldom available above 30 µm.

Overall, the case for using irregular media for preparative chromatography is particulary strong, although we can expect developments in both types of media which will keep the debate alive.

CONCLUSIONS

Flexibility in choice of particle size, pore size and pore volume is needed to provide maximum throughput across a range of separation problems.

Porous irregular silicas are well matched to the requirement of preparative chromatography owing to:

— high capacity,
  30 Å media for small molecules.
  high pore volume media for a general increase in surface area.
— flexibility in pore size, particle size and pore volume
— availability in production-scale quantities.

Overall, they provide a major contribution to the development of preparative chromatography as an important unit process operation.

REFERENCES

[1] *Separation Science and Technology*, Vol. **22** (1987).
[2] M. Verzele and C. Dewaele, *Preparative High Performance Liquid Chromatography. A Practical Guideline*. TEC Gent, Belgium (1986).
[3] J. C. Giddings, *Dynamics of Chromatography*, E. Arnold (1965).
[4] H. Darcy, *Les Fontaines Puliques de la ville de Dijon*, Dalmont, Paris (1856).
[5] I. Halasz and K. Martin, *Angew. Chem. Int. Ed. Engl.* **17**, 901–908.
[6] J. H. Knox and H. M. Pyper, *J. Chromatogr.* **363**, 1 (1986).
[7] P. Gareil and R. Rosset, *Sep. Sci. and Tech.* **22** (8–10), 1953 (1987).
[8] J. Newburger, L. Liebes, H. Colin and G. Guichon, *Sep. Sci and Tech.* **22** (8–10), 1933 (1987).
[9] D. E. Nettleton, *J. Liq. Chromatogr.* **4**, 141 (1981).
[10] E. Godbille and P. Devaux, *J. Chromatogr.* **122**, 317 (1976).

[11] Personal communication.
[12] S. Golshan-Shirazi and G. Guichon, *J. Chromatogr.* **517**, 229 (1990).
[13] M. Verzele and C. Dewaele, *J. Chromatogr.* **391**, 111–118 (1987).
[14] J. Kohler, D. B. Chase, R. D. Farlee, A. J. Vega and J. J. Kirkland, *J. Chromatogr.* **353**, 275–305 (1986).

5

Column technology and packing materials in industrial preparative chromatography

Henri Colin
Prochrom, Champigneulles, France

Modern preparative liquid chromatography offers unique advantages as a production tool in the pharmaceutical industry. The success of this type of chromatography is strongly related to the equipment used (basically the column and the hardware around it, but only the column will be discussed here) and the packing material.

It is clear that the use of columns with a large internal diameter creates specific problems. The most important ones are: the packing and unpacking processes (because, for obvious cost considerations the column hardware has to be reused), the stability of the bed, the convenience of use, the cost of operation, etc. Another important issue (directly related to the column design) is the scalability (that is extrapolating to a large-size column data obtained with a small-size one). This problem can be solved very easily provided the column design is adequate.

Although some personal and emotional considerations are also involved in the selection of the proper equipment, the final choice results from a compromise between the various points previously mentioned. It must also be noted that the best equipment for a given application at a given site may not necessarily be the same as the one for a different application somewhere else.

If the hardware is important, the packing material also plays a very critical role. There are two aspects associated with the packing material: the chemical aspect and the physical/mechanical aspect. The chemical aspect is related to the retention process (absolute retention and selectivity) and concerns the homogeneity of the surface (types of retention sites, presence of active sites, batch-to-batch reproducibility, stability, etc.). Although important, this will not be discussed in this chapter. The physical/mechanical quality deals with the mechanical resistance of the particles, the efficiency (in terms of plate number) of the bed and such aspects as the particle shape, size and size distribution (and thus the pressure drop in the column), pore size and size distribution, etc. Some of these points are discussed below.

TECHNOLOGY OF LARGE-SIZE COLUMNS

The choice of the equipment, particularly at large scale, is an important mattter. The column technology is one aspect of the problem (probably the major one), but there are other ones such as coonvenience of use, price, cost of operation, availability, etc. These aspects will not be included in the following discussion. In the next sections some problems common to any large-size column will be examined, then the different columns technologies available will be addressed as well as the scale-up problem. Also comments on prepacked and user-packed columns will be made.

Note. The following discussion concerns columns with an internal diameter of 2″ and above. Below that size, the columns can be considered as 'oversized' analytical columns. Very often they are used with analytical equipment and their operation does not generate particular problems. A rather large number of $\frac{1}{2}''$ and 1″ columns are available and most of them give satisfactory results. These columns are basically restricted to laboratory applications and/or very small productions (10–20 g/day).

Similarly, this chapter only deals with medium/high pressure (pressure rating above 35 bar — 500 psi) stainless steel columns. Low-pressure columns (often made of glass or some polymers) are not reviewed.

General problems

Some technical problems are common to all large-size columns, whatever their type. These problems concern the column construction, the distribution of the liquid in the column, the packing and unpacking processes and the bed stability.

End fittings

Whatever their type, large-size columns cannot use compression-type fittings as is done with analytical columns (these fittings are available up to 2″, but they are not really convenient to use at such a size). Accordingly, columns must be equipped with flanges. Various designs are available but the principle is always the same. The opening and closing of the column can be a long and tedious operation, particularly at large size since the normal way to connect two flanges together is to use bolts. A large number of them are required at large size (particularly for high-pressure equipment) and these bolts must be tightened with a torque-wrench to get proper sealing. An interesting approach is to use a special clamp system (a type of chain installed around the flanges and holding them together, see Fig. 5.1). Columns equipped with this device are available at small, medium and large size and can be used at high pressures.

Wall thickness

In order for the column to be used at elevated pressure, it is necessary to use thick walls. The thickness depends on the material used to make the column, the column internal diameter and the pressure rating. A safety factor has to be used in the design of the column. Some typical numbers are given in Table 5.1. They are valid for 316 L stainless steel with a pressure rating of 1000 psi.

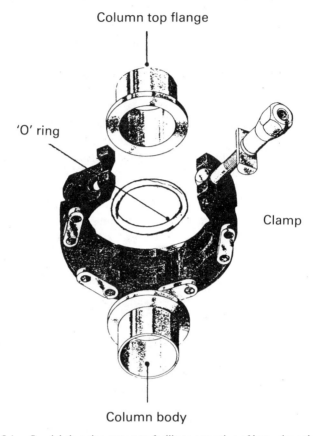

Fig. 5.1 — Special clamping system to facilitate operation of large size columns.

Flow distribution
A proper distribution of the flow of liquid at the column inlet and outlet is of utmost importance. Indeed, in order to avoid band distortion and unsymmetrical peaks, it is necessary to get a 'plug-flow' situation: the liquid velocity must be constant all over the column cross-section. This requires a careful design of the column ends, otherwise the inertia of the solvent molecules entering the column through the solvent inlet line would tend to make the liquid go straight in the column. The molecules would then diffuse only slowly to the wall, creating a curved velocity profile. Properly spreading the flow becomes more and more difficult at increasing column diameter because the ratio of the cross-sections of the column to the cross-section of the inlet line increases. For instance, with a 45 cm internal diameter column, this ratio is about 4000. In order to achieve the plug-flow situation, the design of the column inlet must be such that the velocity of the solvent is reduced by 4000 over a very short distance. This problem is illustrated in Fig. 5.2.

Packing and unpacking process
Analytical columns are usually slurry packed at high pressure (up to 10 000 psi). Although the intricacies of the packing process are not yet clearly understood, it is

Column technology and packing materials

Table 5.1

Internal diameter (cm)	Wall thickness (mm)
50	7.5
60	10
80	11
110	15
150	26
300	28
450	40

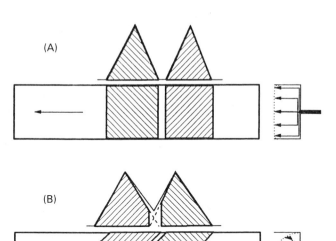

Fig. 5.2 — (A) Poor design of the column inlet providing a non plug-flow situation. The velocity profile is curved, thus distorting sample bands. (B) A plug-flow situation is obtained. The velocity profile is flat and elution peaks are not distorted.

generally considered that a high packing pressure is necessary in order to get a high velocity of the particles impinging on the bed so that they pack tightly. Such a packing procedure is not applicable to columns having a large diameter, because it would require very expensive hardware (not only the column itself, but also the pumping system and the slurry reservoir).

It is thus necessary above a certain size to use a column technology that allows packing at a pressure equal or less than the pressure rating of the column. There are several solutions to this problem.

One solution is the scram approach. It consists in connecting the column to a slurry reservoir and using a floating piston (made of teflon) to force the slurry into the column by applying a liquid pressure on top of the piston. This technique is applicable for small/moderate size columns but cannot be used at large size. Moreover, it does solve the problem of the bed stability (see later).

Another possibility is to use a cartridge. This is discussed in another section.

Finally, it is possible to use a self-packing column (user-packed) based on a compression technique. This approach is probably the most popular, particularly at large size. It is discussed later on.

The unpacking of a large-size column is necessary for obvious cost reasons (with, maybe, the exception of cartridges). This operation can be extremely difficult, particularly when the column is packed with non-spherical silica materials (non-bonded). There exist dilettant effects which make the bed behave as a piece of rock. Opening one flange of the column and pumping solvent at high velocity at the other end does not produce any unpacking at all. It is then necessary to use a mechanical device to push the particles out of the column. The axial-compression approach seems to be the only way to solve this problem easily (see below).

Bed stability

The stability of a column during time depends on many parameters which can be divided into two types: chemical and mechanical. The chemical type concerns destruction of the material by the solvent and/or the sample. The mechanical type is related to reorganization of the particles in the bed and their mechanical destruction by the shear forces resulting from the flow of liquid through the bed. The column technology may also be responsible for the destruction of the particles (as with axial compression when the compression pressure is too excessive).

Of all the above-mentioned parameters, the reorganization of the particles is probably the major one at increasing column size. This phenomenon can be simply explained as follows. The goal of the packing process is to put the particles as closely as possible to each other in order to obtain a compact structure without void. However, such a perfect piling-up of particles cannot be obtained in practice and there are always regions inside the bed where the particles are rather 'lousily' packed. These regions are unstable and, because of the shear forces created by the flow of liquid, they will eventually collapse. This collapsing will produce a settling of the whole bed and the formation of a cavity at the column inlet (mostly). Diffusion effects will take place in this void resulting in band distortion and loss of separation power.

In an analytical column, this collapsing is slowed down by the presence of the column wall against which the particles can lean and the whole bed can get support (stabilization). With increasing column diameter, two difficulties arise. First, the packing operation becomes more difficult and it is likely that there are more unstable regions in the bed. Second, the wall is further and further away from the centre of the column which then cannot get support from it. This situation is illustrated in Fig. 5.3.

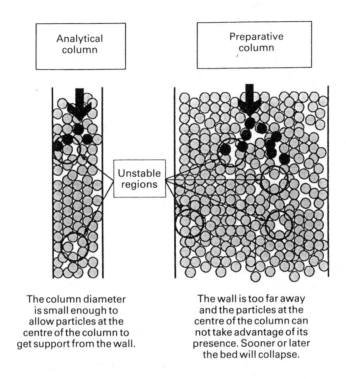

Fig. 5.3 — Bed stability and wall support. At large column diameter, the centre of the bed cannot get support from the wall and eventually collapses.

The technology of large-size columns must be such that the formation of a void is impossible (this does not mean that reorganization of the bed will not happen, but the consequence of this reorganization will be eliminated). This is the reason for the development of compression techniques.

Note. The situation previously described is less critical when packing materials with a large particle size are used. In this case, the column efficiency is (very) small, and the adverse effects of the void tend to disappear. This does not mean that the phenomenon does not take place. It just becomes 'diluted' by the large band-broadening effect created by the large particles.

Column technology

There are basically two technologies for large columns: the regular one (i.e. the same design as analytical columns) and the compression one (there are different types of compression: axial, radial, mixture, dynamic and static). This will be discussed in the following sections. Whatever the technology used, it is also possible to distinguish between prepacked columns and user-packed ones. Finally, cartridges must also be mentioned. They are a special type of column which require the use of a compression technique. Cartridges have some advantages and drawbacks as will be discussed in a following section.

Empty or prepacked columns?
Whereas today most of the analytical columns are brought prepacked, the situation is different with medium/large size preparative columns. There are several reasons for this.

One reason is that large/very large prepacked columns are not available. Except for a few occasions, it can be considered that prepacked columns are only available up to 20 cm internal diameter. Larger columns must then be user-packed, and columns equipped with the technology of dynamic axial compression seem to be the only alternative today.

Prepacked columns (as well as cartridges) are easy to use and require less work from the user. This is true as long as the operator simply has to exchange the column/cartridge by a new one when the separation power of the column has deteriorated (because of bed settling or any other reason). It is always easier, in the case of a problem, to send the column back to the manufacturer and get a new one than to fix a damaged one. On some occasions also, prepacked columns make it easier to work according to GMP rules and do not expose the operator to contact with possibly contaminated packing materials.

Prepacked columns, however, have several drawbacks. First, it is not always possible to find a column packed with the most suitable material for the purification contemplated. This is an important factor for the economics of the purification, particularly at large scale. Indeed, if in many cases several packing materials can do the job at the analytical level, there may be, for instance, major differences between them as far as time of purification and loadability are concerned. Doubling the loadability almost halves the cost of a purification. Another problem with prepacked columns (unless they are equipped with an adjustable head or another sort of compression system) is the stability of the bed, as previously discussed. In some (many?) cases, the possibility of returning a damaged column to the manufacturer is only wishful thinking. Getting back the fixed column may take a long time, which is not really acceptable in production conditions (down-times are expensive). Having several large columns in stock in order to replace on the spot a 'sick' column by a fresh one is usually not economically acceptable. The best solution, particularly for production work, is to be in complete control of the purification column and pack/unpack it at any time, with the material of choice.

Regular-type columns
It is often considered that the maximum size for a regular column (to be packed with relatively small particle size materials, i.e. 10 to 20 µm) is about 5 cm (2″). Regular columns with an internal diameter larger than 5 cm tend to be unstable because of the bed reorganization as previously discussed. It is thus necessary to use a way to

eliminate the void. This can be a movable head which is adjusted when required (static compression) or a dynamic compression technique.

Regular-type columns only differ from analytical ones in the type of end fittings and the wall thickness. They can be obtained empty or prepacked.

Cartridges

Cartridges are basically columns with a flexible wall. In order to be stable and efficient (in terms of reduced plate height), cartridges need to be compressed. The most popular design is based on radial compression (see later). Cartridges have attractive features: relatively low cost, small size of the equipment and convenience of use (see previous section). They also have drawbacks: they do not seem as yet to be available at large size, they have to be purchased prepacked and they are not available with a large variety of media. Very often, also, their pressure rating is rather low and the life time of the cartridge itself may be short (cracks develop after a small number of compression/decompression cycles). In addition, the efficiency of only radial compression at medium/large scale has not really been demonstrated, particularly when small particles are considered. Based on the information available (but there is not much, unfortunately), it seems that cartridges are primarily used in the size range 2.5 to 10 cm (1" to 4").

Compression columns

There are several types of compression column: axial, radial, mixed, static and dynamic. All these technologies have in common the characteristic of using a compression method to keep the integrity of the bed (eliminate voids) and preserve the separation power of the column.

Static and dynamic compression

The compression is static when it is only temporarily applied on the bed. It is dynamic when it is continuously applied. An example of static compression is a column with an adjustable head (using a screw device, for instance). Radial compression columns are of the dynamic type. Dynamic compression is preferable too static compression because the compression is always active, even in case of severe bed shrinkage. With static compression, the compression becomes inactive when the bed has shrunk a little. Although a static compression is a definite advantage over no compression at all, a dynamic compression is a better approach (although usually more expensive). This is illustrated in Fig. 5.4 where data obtained with a 50 mm i.d. column (dynamic compression) packed with 10 μm particles are shown. In the first case (Fig. 5.4, top), after the column was packed (bed length about 25 cm), the pressure was always maintained on the bed (dynamic compression) while solvent flowed through the column (the piston was used to both pack the column and keep it under pressure). A slow upward movement of the piston was seen during time, indicating reorganization of the particles. The stabilization process took about 10 to 15 hours in the conditions of this experiment. During this time the column efficiency is almost constant. In the second case (Fig. 5.4, bottom), after the column was packed, the piston was mechanically blocked and no dynamic compression could take place. In about one day the efficiency dropped from 45 000 plates/metre down to 10 000 plates/metre.

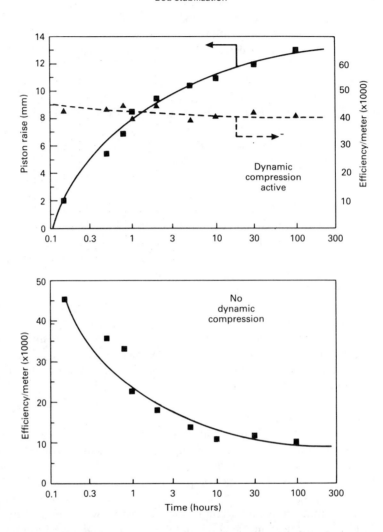

Fig. 5.4 — (Top) The dynamic compression is maintained during column operation. Although the bed slowly reorganizes, the column efficiency is constant. (Bottom) In this case, the dynamic compression is not maintained after the column is packed. The bed reorganization quickly destroys the column efficiency.

Radial compression
The principle of radial compression (Waters–Millipore patent) is shown in Fig. 5.5(A). The column is a cartridge with a flexible wall immersed in a dedicated fluid contained in a stainless steel reservoir. Upon pressurization of this fluid, the column wall is radially compressed, thus pushing the particles together. Such columns packed with small particle-size materials are available at small size, but apparently not at large size. Questions can be raised concerning the feasibility of such an

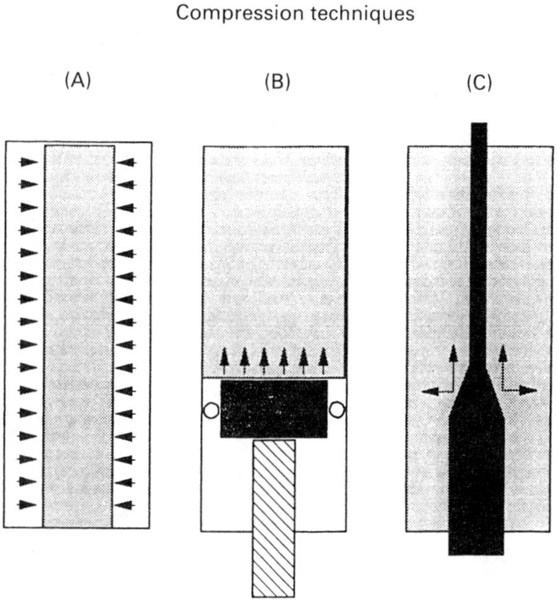

Fig. 5.5 — (A) Radial compression (Waters–Millipore). (B) Axial compression (Prochrom). (C) Mixed compression (Separation Technology).

approach at medium/large size. Indeed, it is possible that, owing to the compression principle, a crust is formed along the internal wall of the column. This crust could very well act like a shell and protect the centre of the bed, thus preventing the compression from extending all over the cross-section of the column.

Axial compression
The principle of (dynamic) axial compression (Roussel–Uclaf patent) is shown in Figs 5(B). The column (made of stainless steel) is packed by the operator. It contains a piston equipped with a special seal. The piston is connected to a dedicated system (i.e. a hydraulic jack) in order to make it go up and down. During operation, the pressure is always maintained on the piston, thus providing the formation of voids. The piston is thus used to pack the column and keep the bed under pressure. It is also used to unpack the column.

The packing is a straightforward operation. It simply consists of preparing a slurry of the desired material in an appropriate liquid, transfer the slurry in the column, close it and raise the piston. It is important to note that the length of the bed can be easily adjusted (it only depends upon the amount of packing material used to prepare the slurry) in order to optimize the column efficiency.

Columns of this type are available up to 80 cm internal diameter. They are probably the most popular technology today. Different manufacturers have developed slightly different designs, but the principle is always the same.

Mixed compression (or annular expansion)
The principle (shown in Fig. 5.5(C)) combines axial and radial compression (Separation Technology patent). The column (made of stainless steel) contains a spindle with a conical shape. Moving this spindle upward produces both radial and axial forces. Whereas this principle is theoretically attractive, it does not seem to be usable at medium/large scale. There is not enough data available to comment more on this approach.

Scale-up

For many reasons it is not convenient to optimize a preparative separation with a large-size column (the one that will be used for the purification process). It is much better to make this optimization work with a small-size column ('analytical' size) and then scale-up to the size required for the preparative process. This scale-up operation can be difficult if parameters other than the column diameter are changed, particularly the number of theoretical plates. Indeed, as theoretically and practically shown, the load that can be put on the column as well as the collection points, depends on the column efficiency. Clearly enough, the chemistry must also not be changed when going from the small to the large column. Accordingly, the best way (and the simplest one) to scale-up is to use in both the small and the large columns the very same packing material (not a different batch or a different particle size) and solvent and keep the bed length, the solvent velocity and the efficiency identical. In these conditions, the scale-up factor is simply the ratio of the column's cross-sections.

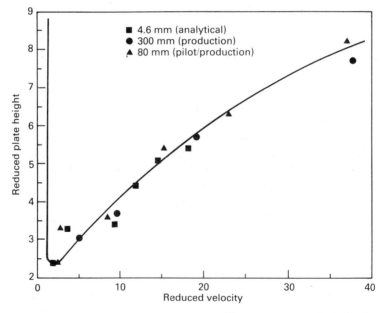

Fig. 5.6 — Reduced plate height data for columns of different diameter packed with the same material (reversed phase 12–45 μm irregular). The 4.6 mm i.d. column is of regular design. The 80 and 300 mm columns are DAC type (dynamic axial compression).

At the equipment level, a successful and simple scale-up then boils down to keeping the same efficiency while increasing the column diameter. The data shown in Fig. 5.6 illustrated results obtained with a 4.6 mm i.d. column (regular type) and two dynamic axial-compression columns (80 and 300 mm i.d.) packed with the same material. As can be seen, all experimental points fall on the same curve, indicating that the column efficiency is independent of the column diameter.

PACKING MATERIALS

Itt used to be considered that preparative chromatography should be done with low-quality packing materials, particularly as far as particle size and size distribution were concerned. This is certainly one of the reasons which have slowed down the use of preparative LC in the industry. It has been only rather recently that people have realized the benefits of using good packing materials. It is also rather recently that such materials have become available in large quantities.

Besides the chemical quality of the packing materials (this is clearly a major conceern but this topic is not addressed here), particle size, size distribution, shape and mechanical resistance are of utmost importance.

Particle size distribution

It is often said that with materials having a large size distribution, the efficiency is controlled by the large particles, and the column permeability (or the pressure drop in the column) by the small ones. This is a bad situation, particularly for large-size equipment. Indeed, large preparative columns having a high pressure rating are expensive (the amount of stainless steeel required, machining, etc). The rest of the equipment (valves, pumps, etc.) is also expensive, particularly if it has to combine large flow rates and high pressures. Consequently, in order to use the equipment efficiently (money-wise), it is necessary to use its full pressure capabilities. This is not the same situation with analytical HPLC where the maximum pressure capabilities of the equipment are almost never used.

All other parameters being constant, the production throughput is proportional to the solvent flow rate; and, at given column pressure, the flow rate is inversely proportional to the column permeability. Consequently, the production throughput is proportional to the column permeability. In other words, a packing material with a large size distribution has to be avoided because it gives at the same time a low efficiency and a low permeability. The presence of very small particles is especially harmful since the permeability is related to the square of the particle size.

If one considers the overall economics of a purification, the savings that can be made by selecting a packing material with a large size distribution compared to one with a narrow size distribution are often ruined by the decrease in production throughput and the increase in purification costs. Indeed, it must be kept in mind that if two packing materials give the same column efficiency and the same chemistry (same selectivity and same loadability) but one is two times more permeable than the other, then the production throughput will be two times smaller with the low permeability material and the purification costs will be much higher (typically between one and a half and two times more). Many cost studies have reached the conclusion that low-quality packing materials are actually more expensive to use than high-quality ones. This is also without mentioning practical problems related to clogging of the equipment by fines, contamination of collected fractions, etc.

Particle size

The choice of the 'optimum' particle size is a difficult issue, although there is a clear tendency today to use particles in the range 10 to 20 μm (a 'typical' material being 15±5 μm). Theoretical work has shown that the optimum particle size depends on the conditions of the purification: not only the selectivity but also the purity, the recovery ratio, the pressure rating of the equipment, etc. The price of the material is also a critical aspect. It should be pointed out that the optimum choice in terms of production (i.e. producing the maximum quantity in a given time with a given equipment) is not necessarily the optimum choice in terms of cost of the purification. This makes optimization of preparative LC at the production level very difficult. It is quite unlikely that very small particles (i.e. 5 μm or less) will be used in preparative LC (at medium/large scale). It is also quite unlikely that new applications will be developed with large/very large particles (i.e. larger than 50 μm). A 'good' size is probably in the range 10 to 20 μm.

Particle shape

The shape of the particles is also a critical point (or at least it seems to be). It is often claimed that, at given size, spherical particles give better results (more plates and less pressure) than angular ones. This is probably not true, but it raises the question of defining the size of an irregular particle. It is very difficult to compare spherical and non-spherical particles. The manufacturing process of these materials is different. Whereas spherical particles are made directly at the right size (the size distribution is usually rather narrow), angular particles are obtained by crushing and sieving big chunks. Accordingly, such particles (almost) always have a larger size distribution than spherical ones. As a consequence, for a given 'average' size (whatever this really mens), angular materials give more pressure. This is not good, as already discussed. However, it is important to mention that the lower permeability does not come from the particle shape itself, but rather from the manufacturing process. It is certainly more expensive to produce spherical particles than angular ones, but sieving irregular particles to obtain a narrow size distribution is very difficult (and time consuming). It may very well be that spherical and angular particles with the same average size and size distribution cost about the same and give similar results.

If the shape of the particle does not thus appear to be a critical issue, it must be pointed out that very often angular materials tend to produce fines. This is just because sharp corners or dents at the surface of the particles are broken during the packing process (whatever the column technology is) or during column operation (shear forces created by the flow of solvent). These fines create troubles, as already mentioned. It is known that very often spherical particles give more trouble-free operation than angular ones. They are also more expensive.

The previous discussion should not lead to the conclusion that irregular materials must be ruled out. Very often, they offer a more attractive quality/price ratio than spherical materials. The choice depends in fact on what the user wants to achieve.

There are basically two extreme situations. In the first case, the material is used for many different products, and/or the particles have to be unpacked/repacked several times. In such a case, it is well justified to spend money on a very good material with a long life time. The opposite situation is when the material is used for a specific application and has to be thrown away after a small number of injections. Then, it may be economically more profitable to spend less money on the packing material. Many intermediate situations can be foreseen.

CONCLUSIONS

As can be seen, the proper choice of the equipment and the packing material is not a simple task. Many parameters have to be considered, some of them being only scientific, others being economical and others also being subjective and emotional.

In any case, it is always a mistake to 'just look at the tree and not see the forest'. Savings made on low-quality equipment and packing material often turn into larger expenditures on a medium/long term basis. This is particularly true when production is concerned.

6

The development of preparative and process chromatographic methods

Derek A. Hill
The Wellcome Foundation Ltd, Dartford, UK

The term 'Preparative Liquid Chromatography' describes any liquid chromatographic separation which is used to isolate, enrich or purify one or more components of a mixture for some subsequent operation, e.g. elucidation of structure, for use as an analytical reference standard, as an intermediate in a chemical synthesis or for sale as a commercial product.

Various authors have attempted to sub-divide the technique, usually on the basis of scale of operation. Bidlingmeyer [1] uses micro-preparative, preparative and macro-preparative, Verzele [2] defines five types, whilst others maay refer to semi-micro preparative (or even semipreparative) as distinct from preparative. The differences are rather arbitrary since with modern instrumentation and, in particular, modern column technology there is little difference in the chromatography between working with a few milligrams on a narrow-bore column or several kilograms on a wide-bore column, although the latter may present different, non-chromatographic, problems such as product isolation or solvent recovery.

More important, however, is the distinction between 'Process Chromatography' and other preparative liquid chromatographic separations. Where chromatography is used as part of a manufacturing process, irrespective of the annual production requirements, it will need to be fully optimized on the basis of minimizing the unit cost of the product. Compare this to other preparative separations, where the time required to optimize the separations may be longer than the time required to complete the separation using a non-optimized or partially-optimized method.

NON-PROCESS PREPARATIVE SEPARATIONS

In industry, as opposed to Academia, the use of preparative chromatography in non-process research and development applications usually results in an indirect cost benefit. Its use may release resources for other work or it may save time in the early stages of a project thus allowing a product to reach the market sooner and with a longer patent life remaining or allowing the project to be dropped sooner. The actual cost of carrying out the separation may be high, but is insignificant in overall terms. The amount of material required will normally determine the amount of optimization work which needs to be carried out. If the requirement is for a few milligrams of product a simple scale-up of an analytical HPLC method to a larger diameter column, possibly coupled to an increased column load along the lines suggested by Knox [3] in the 'Touching Bands' approach will often give the required material very rapidly. When larger amounts of material are required, it will often be better to partially optimize the separation since the time saved in carrying out the separation will outweigh the time taken to carry out the optimization work. In particular, optimization of the method with respect to the ease of recovery of the sample from column eluate may be worthwhile.

The chromatographer must use his judgement and experience to determine the best approach to a particular problem. The situation may of course be different in academic research since budgetary restraints often reduce the availability of modern instrumentation and materials, and a scientist's time is not costed in the same way as it is in industry. In this case optimization may be required for most separations that are carried out.

PROCESS CHROMATOGRAPHY

The isolation and purification of materials in a production environment must be considered not in isolation, but as an integral part of the production process. The decision to incorporate a chromatographic step will, provided that alternative purification methods are feasible, be taken solely on the basis of its cost effectiveness compared to the available alternatives as determined by the unit cost of the product. This is distinct from the use of preparative chromatography in research and development where its use may often be justified on the basis that the saving of time can be more important than product cost. In considering synthetic products as opposed to natural products, the quality of the crude product can be affected by the choice of synthetic route or the conditions used in a reaction. It may therefore be possible to modify the quality of the crude product in order to simplify the chromatographic purification. Consider the following example:

$$A \rightarrow B + \text{side products}$$

If the reaction is continued until all of compound A has been consumed, the purification step may require the isolation of the product B from a complex mixture. However, stopping the reaction before appreciable amounts of side products are formed may allow a comparatively easy separation of A and B to be carried out, with the starting material subsequently being recycled. The overall economics of the process may well be improved since the throughputs in the expensive chromatographic step would be considerably higher. In the case of a product which may be prepared *via* different synthetic pathways, the impurity profile of the crude product may well depend upon the particular synthesis used, and a route which produces a lower yield may, in fact, be cost effective if the impurity profile of product produced *via* the route allows the chromatographic step to be simplified with a consequent reduction in cost. A second example is the use of a combination of purification techniques, such as suggested by Jones [4] where he advocated the use of chromatography to recover valuable material from recrystallization mother liquors.

Consider the situation where a product may be purified by chromatography or by recrystallization. If the value of the product is high enough, the higher yields obtained using a chromatographic purification will outweigh the lower processing costs of a recrystallization and will, therefore, be cost-effective. However, the use of a combination of both may well prove to be the most economical. Thus, consider the example of a crude product with a value of $1000 \, kg^{-1}$. If this material is purified by recrystallization in 80% yield with a processing cost of $10 \, kg^{-1}$ the cost of the purified product will be $1260 \, kg^{-1}$ whilst, if a chromatographic purification is used at a cost of $100 \, kg^{-1}$ and in 95% yield, the purified product cost will be $1153 \, kg^{-1}$, i.e. chromatography is cost effective. However, if the material is recrystallized and chromatography is used to recover material from the recrystallization mother liquors, allowing for a higher cost of the chromatography ($200 \, kg^{-1}$) because of the crude nature of the material in the liquors, the cost of the product will be reduced further to $1103 \, kg^{-1}$. An alternative approach would be to use chromatography on the recrystallization mother liquors, not to produce pure product, but to enrich its quality to that of the crude product and subsequently to use recrystallization to obtain pure material. The cost of the product would be similar to that of material produced by the recrystallization/chromatography process, but in this case all of the product would undergo an identical final processing step (recrystallization). This might be advantageous in, for example, the registration of a drug substance with the Regulatory Authorities.

These examples demonstrate the need for the process chromatographer to get involved with process development at an early stage, and to educate the development chemists and engineers to the idea that process chromatography is a viable alternative to be considered in the early stages of a programme of work and not as a last resort. We may even avoid the common problem of the chromatographer being presented with a bucket of black tar and the request 'Please purify this by Yesterday'. The situation in the field of natural products as opposed to synthetic products is of course different since, usually, the requirement is to isolate one or more (often minor) components from a complex matrix, and chromatography has long been recognized as the best approach to this problem.

STRATEGY FOR OPTIMIZATION IN PROCESS CHROMATOGRAPHY

The starting point for optimization of a chromatographic separation must be a definition of the quality of product required from the separation. If the purified product is to be marketed directly it may need to assay at >99.5%, but in the case of material which is an intermediate for further chemical elaboration, 95% or even 90% may well be adequate. The importance of this point cannot be over-stressed since the higher the quality of product which is required, the lower will be the throughput and the higher will be the cost. It will also be useful to have some estimate of the eventual annual production requirements, and of the amount of material which will be required during the development stage of the project since this will have to be produced as a part of the process development work. Assuming that we already have some knowledge of the chromatographic behaviour of a particular material from analytical work, how then do we tackle the problem of devising a fully optimized process for its isolation and/or purification? It is tempting to take the analytical system and to use it for the basis of the optimization work. However, in using this approach there is considerable risk of producing processes which, although fully optimized in local terms, are not optimized on a global basis. For example, a method might be optimized using an aqueous mobile phase on a bonded-phase silica when a non-aqueous mobile phase on silica would have given higher throughputs and lower cost, or *vice versa*.

Unfortunately, in addition to developing a process there is usually a demand for material to be produced (often as soon as possible) and this can mean that we concentrate our efforts around the systems with which we are familiar, i.e. the analytical system, to produce initial supplies of material with the intention of returning as time allows to reinvestigate the separation. One must be aware that there are problems associated with this approach. For instance, if the material to be produced will be used for toxicity studies on a potential drug substance, a change in the purification process will often necessitate a repetition of the toxicity study since its impurity profile is likely to be altered. The final choice of purification method should therefore be made as early as possible in the development stage of the project.

A good point at which to start the development of the chromatographic process is an investigation of the solubility of the product in a variety of solvents. There is little point in developing a system which gives superb analytical resolution but where lack of solubility of the product in the mobile phase will only allow very low throughputs to be achieved. It is also advisable at this stage to check the stability of the product in solution, bearing in mind that the product will have to survive an isolation procedure once the chromatography has been completed. It is worth thinking ahead at this stage to the product isolation step since it may affect the choice of chromatographic method. Consider the example of an aqueous mobile phase with an organic modifier. If the product is insoluble in water, evaporation of the modifier will allow product recovery by filtration, but, when the product is water soluble, removal of the water may require lyophillization and the consequent installation of expensive plant.

It has been demonstrated [5] that, on a cost basis, there is little difference between using bonded-phase silicas with aqueous mobile phases and silica with non-aqueous solvents since the higher cost of the bonded-phase silica is balanced by the lower cost of solvents and the increased lifetime of the column. The cost of the isolation procedure may therefore, in some cases, by the major factor influencing the choice of method. The next step is an examination of the chromatography of the material on a variety of stationary phases and with a variety of mobile phases. It is useful to use TLC in addition to HPLC at this stage, since it can be used as a very fast way of scanning a large number of potential systems to obtain a rough guide to chromatographic behaviour. It will also demonstrate the presence of any strongly retained impurities which could lead to reduced column lifetimes.

By this time, the chromatographer should have developed a 'feel' for the compounds with which he is working and should be able to use his experience to decide which avenues are worthy of further exploration. In some cases the scope for development will have been considerably limited by the investigations already carried out, but in other cases there will be numerous avenues of investigation still open to us. Tiselius [6] defined three modes of operation of chromatographic column, namely elution, displacement and frontal. The use of the elution mode in preparative chromatography is well documented, but process examples are more difficult to find sice these are industrial applications and therefore remain unpublished for reasons of confidentiality. Displacement chromatography as a preparative method has been revived by Horvath *et al.* [7] and Verzele *et al.* [8] among others, and more recently [9] frontal chromatography has been found to have preparative utility. To these we can add the 'sample self-displacement technique' which has been developed by Newburger and Guiochon [10] and can sub-divide elution chromatography into isocratic and gradient. The problem for the chromatographer trying to optimize a process is to decide which of these methods will be cost effective when applied to the particular example with which he is working. Katti *et al.* [11] have compared the performances of overloaded elution chromatography and of displacement chromatography. The results suggest that neither method stands out as having an advantage over the other, and that it would be necessary to compare both methods for any particular separation problem. Furthermore, since the advantages of one against the other may lie not in the chromatography but in the downstream processing, the comparison must be made on the basis of product cost.

The 'self-displacement' approach might be expected to combine the advantages of the two methods and may become the method of choice now that the basis for inducing the sample to undergo self-displacement is better understood, whilst a frontal approach may be cost effective in some cases and, in particular, where an enrichment or partial purification only is required, since as Guiochon and Katti [12] pointed out the method is not well suited to the production of pure material. When elution chromatography is considered the situation is further complicated. Is, for example, it better to use gradient elution for the separation of a mixture of compounds whose chromatographic behaviour differs widely, or is it preferable to use a two-stage process where a crude separation to produce a group of compounds with a more restricted range of chromatographic behaviour is followed by an isocratic method to isolate the compound of interest? Can we increase throughput by overlapping injections so that, in the column dead time between the injection and the commencement of elution, the product from the previous injection is eluting and the column is thus used to its maximum? With such a wide range of options open to us it is impossible to predict which area is likely to provide the most cost-effective solution to any particular separation problem, and hence in which areas the emphasis of the work should be placed, and it may well be necessary to carry out some optimization work on several different methods before deciding which is likely to providee the optimum process. In these laboratories this work is carried out on either a 25 mm or a

50 mm i.d. column since this scale of operation will enable a study of the isolation of the product from column eluate to be carried out so that the cost of various options can be calculated, and the method which finally will be finely tuned for process use can be selected.

In the last few years, a considerable amount of work has been published on the theoretical aspects of preparative separations. From the original paper of Knox and Pyper [13] methods have been refined and developed by several groups [14–18] so that computer methods can be used to simulate and to optimize preparative separations in such a way as to obtain a fair degree of agreement with practical results. These methods will, without doubt, be of great assistance in process optimization in the not too distant future, but at the present time are probably best left to the theoreticians. However, it is to be recommended that the practical chromatographer follow these developments so that he may take advantage of them as they become more reliable and simpler to use, even though their relevance to his work at this time may seem to be rather limited, particularly since they appear, thus far, to have been applied solely to binary mixtures rather than the multi-component mixtures which are common in 'real-life' situations.

CONCLUSION

Although the use of chromatography in preparative and process applications has increased considerably in the last few years, there is still a tendency for it to be treated as a last-resort technique in a production environment, rather than an alternative to be considered from the initiation of a project. The belief that it is intrinsically an expensive technique or that it would not be possible to use it for the production of tonnage quantities both seem to be fairly widespread among both production and development personnel. The onus therefore lies on us as chromatographers to alter these views so that chromatography may assume its rightful place in the production environment.

REFERENCES

[1] B. A. Bidlingmeyer, *Preparative Liquid Chromatography*, Elsevier, Amsterdam (1987).
[2] M. Verzele and C. Dewaele, *Preparative High Performance Liquid Chromatography Conference*, Gent (1985).
[3] J. Knox, *Guidelines for Developing Preparative HPLC Separations*. Presented at the 7th International Symposium on Preparative Chromatography, Gent (1990).
[4] K. Jones, *Chromatographia*, **25**, 577 (1988).
[5] K. Jones, *Chromatographia*, **25**, 547 (1988).
[6] A. Tiselius, *Ark. Kemi Mineral. Geol.* **16A**, 1 (1943).
[7] C. Horvath, A. Nahum and J. Frenz, *J. Chromatogr.* **218**, 365 (1981).
[8] M. Verzele, C. Dewaele, J. Van Dyck and D. Van Haver, *J. Chromatogr.* **249**, 231 (1982).

[9] D. A. Hill, P. Mace and D. Moore, *J. Chromatogr.* **523**, 11 (1990).
[10] J. Newburger and G. Guiochon, *J. Chromatogr.* **523** 63 (1990).
[11] A. M. Katti, E. V. Dose and G. Guiochon, *J. Chromatogr.* **540**, 1 (1991).
[12] G. Guiochon and A. Katti, *Chromatographia*, **24**, 165 (1987).
[13] J. H. Knox and M. Pyper, *J. Chromatogr.* **363**, 1 (1986).
[14] S. Golshan-Shirazi and G. Guiochon, *J. Chromatogr.* **536**, 57 (1991) and references cited therein.
[15] L. R. Snyder, J. W. Dolan and G. B. Cox, *J. Chromatogr.* **484**, 437 (1989) and references cited therein.
[16] F. D. Anita and C. Horvath, *J. Chromatogr.* **484**, 1 (1989).
[17] J. M. Jacobson and J. Frenz, *J. Chromatogr.* **499**, 5 (1990).
[18] Q. Yu and D. D. Do, *J. Chromatogr.* **538**, 285 (1991).

7

Determining operating parameters in process chromatography

Michael Kelly
St. Nabor, France

The title of this chapter evokes the idea of scaling up, from analytical- to industrial-scale chromatography, in line with my original lecture at the Loughborough University short course on which this chapter is based. To give it such a title would, however, be to yield to a preconception which I wish not only to avoid, but to positively distance myself from: that scaling up involves establishing whether certain predetermined analytical conditions 'work' at the preparative or process scale, or, perhaps, what are the feedstock quantities and column-size requirements for making productive use of an analytical separation. Lectures, seminars, articles and chapters on scaling up chromatography are probably almost as numerous as books and courses on improving your memory or increasing your reading speed, and have something in common with them: they aim to teach you the mechanics of the process, the HOW?, but the important question is WHAT?, not only how to read or remember, but what is worthwile reading or remembering. In the course of this chapter I will deal with the mechanics of scaling up from analytical chromatography to larger scale, as they are an essential practical requirement, but I want to do so from the initial standpoint of considering the requirements of chromatography at the larger scales, and from these determining the range and type of conditions for the optimization process, which has of practical necessity to be carried out at the analytical scale. It is perhaps unfortunate that we use the expression *analytical scale* when we refer to small-scale chromatography on the laboratory bench rather than out in the pilot plant, for *analytical* refers to a purpose, an objective rather than a scale, in the same way that *preparative* scale can refer to milligrams (or less) or kilograms (or more). In setting up the parameters for large-scale chromatography we will therefore have recourse to both analytical chromatography, to monitor the results of the optimization process and the scaling up, and to small-scale chromatography to examine various combinations of stationary and mobile phase, load, flow rate and detector parameters in an efficient and economical manner. A good review of the scaling up of small-scale separations, with particular reference to the 'laboratory-preparative scale' is given by McDonald and Bidlingmeyer [1], with a more detailed discussion of the mechanisms involved than is possible here.

One important aspect this chapter will *not* cover, is the cardinal question of WHAT substrate. Only you can make this decision, and maybe not even then. However, my personal opinion is that the choice of suitable candidates for purification by chromatography is often too limited and too much dictated by the *difficulty* of the separation as an analytical problem, rather than by an evaluation of the most suitable of a variety of possible methods [2]. A common basis for deciding to use chromatography seems to be: the components of this mixture can only be separated by very sophisticated liquid chromatography, therefore no other technique will be adequate. The corollary is, this separation is easy, some other less noble technique should be good enough. The only criterion for deciding to use large-scale chromatography is that it is the best of the methods evaluated, considering the constraints of the operation *at the final operational scale*. Chromatographers often seem to regard crystallization as a simple technique, needing only a beaker and some solvent: consider the equipment necessary at a production scale, and the questions of yield and purity reproducibility. A chromatographic separation may be easy, but chromatography should not be rejected, if all the alternatives are difficult.

The first stage of defining the parameters of a large-scale separation is therefore to consider the characteristics of large-scale, as opposed to small-scale, chromatography. We will consider here the stationary and the mobile phases, and just two aspects of hardware, detector choice and characteristics of gradient-generation hardware, as they relate to the development of the method. I have reviewed the hardware design aspects of process-scale chromatography elsewhere [3].

STATIONARY-PHASE CONSIDERATIONS

Let us first consider the various types of stationary phase, comparing their characteristics for analytical use with those for preparative use. In the choice of class of stationary phase the analytical method, if one exists, will probably be a good guide, as this is often determined by class of compound and acquired experience. However, the type of stationary phases available should always be reviewed at the beginning: changes later are expensive.

Where either of two types of stationary phase can be used, the *large-scale* suitability of the phase, and the mobile phase type it predicates, should be given priority over superiority of analytical result. For example, if both normal and reversed-phase methods are possible, the reversed-phase will normally be referred because of the stability of the stationary phase, and the generally more-convenient type of mobile phase, even if the better analytical results are obtained on straight phase.

Silica
A less-expensive bulk phase than bonded silica-based materials, but the shorter life and the greater expense of mobile phases will generally counteract this advantage. More suitable for self-packing columns than prepacked or cartridge systems, by the labour cost and associated down-time involved should not be forgotten. The choice of mobile-phase solvents is wider than for reversed-phase conditions, but the solvents tend to be either toxic or inflammable, and although easier to recover, are more expensive. For non-polar compounds that are insoluble or unstable in hydroxylated solkvents straight-phase chromatography may be the only option. In this case the bonded straight-phase packings should also be considered.

Reverse phase
Sometimes regarded as synonymous with C-18, the silica-based bonded packings have the advantages of stability, resistance to contamination, cheap and easily-modified mobile phases based on a limited number of solvents, of which water is the chief actor, and of a very wide range of established applications. Further, even within the family of C-18, there are a large number of possibilities for optimization of the process, with variables such as carbon load, pore size and particles size. In the optimization process considerations of cost, availability, reproducibility and ruggedness should weigh heavily.

In quality control chromatography the concept of *master batch* is gaining ground, to provide assurance that the characteristics of an analysis will not be modified by inopportune batch-to-batch variations in column performance. On the process scale, the stationary-phase requirement per column is high (depending of course on the scale of operation) compared to batch sizes, so the master-batch concept will be largely inapplicable. The defence against batch variation is to give priority to ruggedness over critical optimization in method development. If the scale is to be really large and the production duration long, discussion of availability with the stationary-phase manufacturer becomes essential. Other stationary phases in this class should not be forgotten (C_4, C_8, phenyl, etc.), but the question of availability should not be forgotten either.

An interesting stationary-phase development of great potential for preparative and process operations is the porous packings in which the individual particle is traversed by large-diameter 'bulk-transfer' pores providing short-distance, rapid-equilibration access to the narrow 'separation' pores [4,5].

Ion exchange
Considerations of column life, method reproducibility and eluent choice and cost make this probably the most widely used large-scale chromatographic mode. It is of course limited to water-soluble compounds ionizable within the pH range of the ion-exchange material, which may be resin-based or silica- or alumina-based. If it is not given more consideration in this chapter, it is because of my own lack of experience, and the greater expertise of authors of other chapters, than any inadequacy or irrelevance of the method.

Size-exclusion
Widely used for small-scale preparative purification this mode has advantages of reproducibility and predictability, but disadvantages of load and efficiency. The separation is not determined by the eluent (except to the extent that the eluent may cause changes in tertiary structure), but the eluent is determined by considerations of solubility of substrate and stability of the stationary phase, and is thus eminently suited to the principle of determining conditions by considering requirements of the full-scale process. Preparative applications of this technique have been recently reviewed [6].

The confines of this chapter do not permit a thorough review of all aspects of the choice of chromatographic mode and of specific stationary phase: my objective is simple, to pose the question: 'Have I considered the choice of stationary phase, from the standpoint of large-scale operation?'

MOBILE-PHASE CONSIDERATIONS

In determining the mobile phase for a large-scale purification or an analysis, different parameters are of interest. Table 7.1 summarizes some of these differences.

Table 7.1 — Solvent considerations for process chromatography — 1. Importance of various factors in analytical and process applications of HPLC

Parameter	Analytical	Process
Viscosity	High	Moderate
Price	Low	High
Toxicity	Increasing	High
Flash point	Low	High
Boiling point	Low	Moderate
Prior use	Moderate	Low
Selectivity	High	High

The fact that a given solvent is at a disadvantage for a parameter shown as of high importance does not mean that the solvent cannot be envisaged as an eluent component, only that its use will involve measures that will have an effect on the cost of the process. Thus a toxic solvent is not excluded, but before developing a method that tends towards use of this solvent, the precautionary measures likely to be necessary must be considered. These are likely to include ventilation and enclosure, monitoring of workplace air and possibly medical examination of workers, and restrictions in the making up and transporting of the eluent. These being considered, it is probably safer then to work at the large scale than in the laboratory where a variety of toxic solvents may be in use without specific measures. Likewise the use of

Ch. 7]   **Determining operating parameters in process chromatography**   107

a viscous solvent is not excluded — indeed in practice viscous solvents seem to be quite popular — but the consequences for flow rate and back pressure must be considered from the outset. Table 7.2 shows some of the commonly used solvents,

Table 7.2 — Solvent considerations for process chromatography — 2 Solvent parameters

Solventr	TLV (ppm)	Flash pt.	Boiling pt.	Viscosity
n-Hexane	50	$-6°C$	64°C	0.33 cP
n-Heptane	400	$-1°C$	96°C	0.41 cP
Dichloromethane	100	None	40°C	0.45 cP
Trichlorofluoroethane	1000	None	47°C	0.66 cP
Ethyl acetate	400	$+7°C$	77°C	0.46 cP
Acetonitrile	40	$+5°C$	82°C	0.36 cP
Methanol	200	$+11°C$	64°C	0.6 cP
Ethanol	1000	$+12°C$	78°C	1.2 cP
Acetone	750	$-20°C$	55°C	0.36 cP

and some less commonly used ones. This list is not intended to be exhaustive, but rather illustrative. The usefulness of a solvent in terms of chromatographic selectivity cannot be tabulated, as it is a function of the particular separation problem, so the parameters shown in Table 7.2 reflect the suitability of solvents for large-scale chromatographic operations in practice.

The TLV is the threshold limiting value to which workers may be exposed for specified time limits: a higher value represents a less-toxic substance. Each country has its own legislation and its own lists of values, and it is these which will be applicable when the process is implemented. An established facility will already be applying the relevant national legislation, but not necessarily for the solvents proposed for the new operation. For method development any up-to-date list of TLVs will provide the necessary awareness of potential problems [7]. The flash point is an indication of fire or explosion hazard. Again, although the requirements of explosion-proof construction are largely harmonized (the CENELEC standards are applicable in all 19 signatory European countries), the classification of the risk in a specific situation — location, equipment, sources of hazard, other hazardous activities in the area — remains a national responsibility [8]. The use of a low flash-point solvent indicates the need to discuss the process with production and safety management *at the development stage*. The boiling point also refers to safety (both fire and toxicity; it is easier to control the concentration of a higher-boiling solvent) but also to ease of substrate and solvent recovery and extent of thermal degradation. Viscosity is directly related to back pressure, a significant consideration in a larger-scale operation, even what is generally known as lab-prep, because pressure limits of both pumps and columns are generally lower than in small-scale systems, a limit of about 140 bar (2000 psi) being common.

STRAIGHT-PHASE ELUENTS

Looking at the first two solvents, hexane and heptane, we can see that although hexane is probably the more commonly used in analytical applications, there are serious reasons for preferring heptane: it is slightly less inflammable (but not enough to avert the need for explosion-proof systems in production), but above all the toxicity is much less (this surprising fact is due to the metabolism of *n*-hexane to the very toxic hexa-2,5-dione, which is not mirrored by its homologues). In chromatographic performance fluorocarbon-113 (1,1,2-trichloro-1,2,2-trifluoroethane) is comparable to hexane, but is greatly superior in terms of large-scale suitability, with its non-inflammability and its virtually total absence of toxic effects (1000 ppm is the highest TLV accepted). An example of the development of a method for large-scale use from an analytical method using hexane follows later in this chapter. Of course, the Montreal and London protocols require the cessation of use of CFCs from 1997, so to develop a method today which uses this solvent as eluent would be short-sighted. However, the principle of looking at parameters relevant to production when developing a method is still valid, and it will be of interest to examine some of the proposed replacements, although these will almost certainly be more toxic, more polar and less stable (stability being the cause of the environmental problems of CFCs).

REVERSED-PHASE ELUENTS

Looking further down Table 7.2 we can consider the solvents used in eluents for reversed-phase chromatography. Despite an increasing concern for its toxicity, acetonitrile is still undoubtedly the analytical chromatographer's favourite. In many cases it gives much better analytical chromatograms, in part due to its low viscosity and the improvement of mass-transfer, and in part due to the absence of hydroxyl groups to interact with polar analytes. However, its toxicity will necessitate control of the production environment, and also careful monitoring of residues in the product. From a production point of view the lower alcohols will be preferable, and also from the point of view of final product composition [9]. Thus it remains for the user to evaluate each case, but certainly no large-scale purification using acetonitrile should be considered without having examined the chromatography using the lower alcohols methanol, ethanol and *iso*-propanol. The possibility of using acetone is interesting, as it also raises the question of the chromatographic method determining the hardware required. From a solvent viewpoint acetone is attractive, having low viscosity and toxicity, being a readily available bulk solvent with low boiling-point for product recovery (but inflammable!). The reason it is not used is the overwhelming predominance of UV detection in analytical applications. Acetone cannot be used in conjunction with a UV detector because it absorbs strongly in the proportions used as an eluent. However, the advantages of UV detection for analysis (sensitivity and selectivity) are not necessarily translated to the large scale, where the sensitivity has often to be attenuated and the purpose is not to analyse but to provide a 'map' to follow the progress of the production operation. However, the use of acetone will require the use of a refractometer or densitometer, or the inclusioon of one in the specification of the production chromatograph. It can only be used in gradient separations if off-line fraction analysis is used, or if the fraction cut-points can be determined from a knowledge of status of the gradient — acetone provides a good monitor of the gradient composition! I cannot repeat often enough that the object of my examples is not to promote or discourage the use of a particular solvent or method, but to encourage the consideration of the optimum method from a consideration of the requirements of the eventual production environment, which is valid also in general terms for 'lab-prep' though less critical.

A PRACTICAL EXAMPLE

Now let us look at some of the considerations as they apply to a real example, the investigation of the feasibility of separating pilot-scale quantities of diastereoisomers. The first step is the definition of the project. MacDonald and Bidlingmeyer [1] provide a good check-list of important factors to consider in defining the problem. In this case we had:

Material available:	mixture of diastereoisomers in ratio 60:40 with small amounts of impurities. Non-polar compounds insoluble in water. Not heat-labile.
Product required:	each diastereoisomer in >98% isomeric purity and maximum recovery. Removal of other impurities not essential.
Quantity required:	30 kg of mixture purified.
Other conditions:	Time scale 3–4 weeks. No explosion-proof equipment or workplace available.

An analytical HPLC method had been developed.

The first point to notice is that the recovery required is not defined. It is not unusual for the first definition of a project to require '100% purity with 100% recovery' or 'maximum purity with maximum recovery'. Neither is a working definition of a project. It may be that in the course of development the definition must be changed, but the project must start with a defined need. In this case about 90% was thought acceptable.

The analytical method was a straight-phase separation, using a silica column and isocratic hexane/ethyl acetate 90:10 and the chromatogram is shown in Fig. 7.1(a). The detection was by UV absorbance. The selectivity α was 2.05.

Chromatographically this was a good base for a separation. In terms of working conditions it was certainly not. First lesson: a suitable method depends on *where* it is to be applied. The use of hexane may be acceptable in a factory environment where hexane is regularly used and fire and health precautions are in place, but in this case was unthinkable. Likewise a simple change to heptane or octane, which would not be

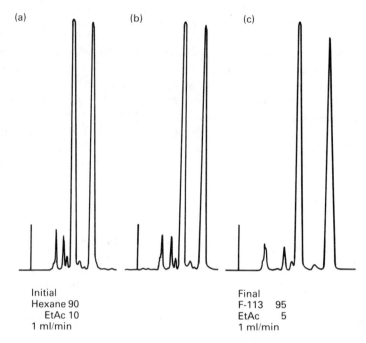

Fig. 7.1 — Method development at small scale. (a) Original analytical conditions, 90% hexane, 10% ethyl acetate, silica column. (b) Intermediate stage, 50% of hexane replaced by trichlorofluoroethane. (c) Final non-toxic, non-inflammable conditions, 95% trichlorotrifluoroethane, 5% ethyl acetate.

expected to change the chromatography significantly might be adequate where an explosion-proof environment was available but the toxic risk of hexane was unacceptable. Of course, if only one set of conditions is found to resolve the problem, it must be used and the working environment adapted to accommodate them: hence the need to cast a wide net in the development phase to reduce the risk is not finding the most suitable conditions.

A number of non-inflammable solvents are available for straight-phase chromatography, mostly chlorinated hydrocarbons with varying degrees of toxicity. At the time of investigation, CFCs were acceptable and it seemed worthwhile investigating their use in this case. The other mobile-phase component was acceptable, and as that is the modifier, with a greater influence on the chromatography than the virtually inert carrier solvent, the first changes were made to the inert solvent. Had the new solvent proved inadequate, a second stage would have been necessary to investigate other modifiers to restore the lost selectivity.

The conditions were modified in a stepwise manner, replacing 10% of hexane by trichlorotrifluoroethane (F-113) in each step, so as to retain identification of the components and their k' values. Elution order, and even retention mechanism, can alter even on changing only the percentage composition of the same solvents in an eluent [10], so care is needed in this stage of development. Reversal of elution order can be very valuable in obtaining good purity and recovery of a second or a minor component: in this present case reversed elution would have been desirable, especially for the component in slightly lower concentration, but was unfortunately not observed. An intermediate composition is shown on Fig. 7.1(b), and the final stage of replacement of all hexane by trichlorotrifluoroethane is shown in Fig. 7.1(c). The ethyl acetate concentration was gradually decreased through the sequence to maintain adequate retention, as trichlorotrifluoroethane is a slightly stronger eluent than hexane. A number of other modifiers were investigated, as an improvement in α is never without interest, but nothing more suitable than ethyl acetate was found. The fact that neither component is hydroxylated augurs well for the life of the silica stationary phase. Suitable conditions for large-scale operation were thus available which maintained adequate chromatographic separation. The next stage was to carry out the scale-up procedure to determine maximum possible load for the purity requirement, and determine whether this would permit the necessary recovery and throughput to meet the quantity and time requirements.

Two steps of the small-scale column loading study are shown in Fig. 7.2. At 10 mg load on a column 4 mm × 15 cm baseline resolution is still obtained and the purity and recovery requirement certainly readily achievable. However, using the equation for scaling-up load:

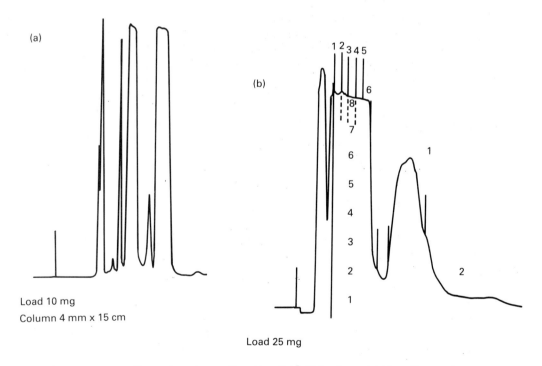

Fig. 7.2 — Loading study on 4 mm × 15 cm NovaPak® silica column. (a) 10 mg diastereoisomer mixture. (b) 25 mg diastereoisomer mixture.

$$\text{Load}_2 = \text{Load}_1 \times (D_2/FD_1)^2 \times (L_2/L_1) \qquad (7.1)$$

where D is column diameter, L is column length and subscripts 1 and 2 refer to the known and the required conditions, respectively.

It can already be appreciated that this load will not be sufficient to fulfil the time and quantity requirements with columns of 10 or even 20 cm in diameter. A load of 25 mg on the small column was therefore tested. Apart from the fact that there is no baseline resolution, the extent of resolution cannot be estimated, as the detector is saturated. Fractions were collected through the major peak as indicated in Fig. 7.2(b) and analysed. The results for two of the fractions are shown in Fig. 7.3 and show that the required purity can be met at this loading.

The purity obtained may seem surprising, but it should be remembered that the detector is saturated and does not in any case distinguish between the two isomers (only a polarimeter detector will achieve this in such a case, as there is unlikely to be any difference in UV spectrum of diastereoisomers sufficient to be useful for photo-diode array detection). It should also be borne in mind that baseline resolution is a requirement for quantitation but not for separation. When evaluating the suitability of a separation for preparative use, the computer simulations published by J. K. Whitesell [11] are useful, and encouraging.

The investigation was then transferred to a larger, preparative-scale column, for two reasons, firstly one of availability and suitability of columns packed with the stationary phase intended for the full-scale operation, and secondly the ability to prepare in the same operation a few grams of pure reference material. The scale-up equation is used to determine the equivalent load for the 5.7×30 cm column, and the volume flow rate needed to give the same linear flow rate calculated from the second scale-up equation:

$$\text{Flow rate}_2 = \text{Flow rate}_1 \times (D_2/D_1)^2 \qquad (7.2)$$

where the symbols have the same significance as in Eq. (7.1).

These two equations give a load of 11.5 grams and a flow rate of 205 ml/min as equivalent to the small-scale parameters of 25 mg and 1 ml/min. It is not of course by chance that these correspond to the capabilities of the preparative and process equipment available and the production rate required, because for a short-term project tailor-made equipment will not be developed. For a study of a manufacturing process designed to define the equipment requirements there will be fewer constraints, although the small-scale phase should consider real-world limitations — for instance, the flow rate and back pressure will be subject to constraints imposed by engineering concerns, even if tailor-made equipment is built.

The preparative-scale separation is shown in Fig. 7.4. Fractions of 100 ml were collected, except two of 50 ml in the cross-over region. Each fraction was analysed after being thoroughly mixed, and a table of isomeric purity and quantity produced, from which the numbers of the fraction to be combined to give two products meeting the composition and recovery criteria can be determined (Table 7.3).

Ch. 7] Determining operating parameters in process chromatography 113

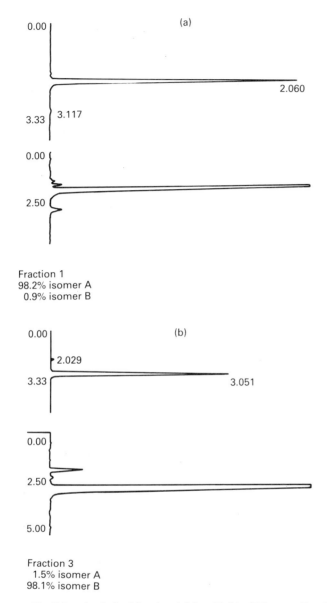

Fig. 7.3 — Analysis of fractions 1 (a) and 3 (b) of 25 mg small-scale separation.

Examples of chromatograms of fractions are shown in Figs 7.5 and 7.6.

From Table 7.3 it can be seen that the criteria for isomer A can be met by combining fractions 2 and 6, but fraction 6 adds little to the recovery and the isomeric purity is better without. As the composition of fraction 6 is close to the composition of initial material, it represents no improvement, and this fraction would be better employed by rechromatography. Likewise, including fraction 7 in the isomer B

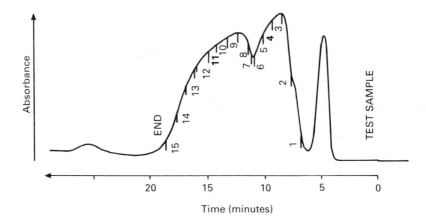

Fig. 7.4 — Preparative separation of diastereoisomer mixture. Load 10 g. Column 5.7×30 cm PrepPak$^{\infty}$ silica.

product reduces the isomeric purity and adds little to recovery. Addition of both these fractions to subsequent runs would increase the recovery of each isomer to about 96%. Including fractions 14 and 15 in isomer B would add little, but increase substantially the cost of recovery, and increase the potential for heat degradation of the product. A synthesis of the fractions combined for each pure isomer is shown in Fig. 7.6.

Applying the scale-up Eq. (7.1) for a process-scale column of 20×60 cm gives a load per run of $10 \times (20/5)^2 \times (60/30) = 320$ g, of which 300 g would be fresh starting material and 20 g re-chromatographed from a preceding run. To separate 30 kg of starting mixture would then require three weeks at seven to eight runs of about 40 minutes each per day, allowing for the extra time needed to remove the late-eluting impurity (fraction 8 in Fig. 7.2(b). Thus in this example the criteria originally stipulated can be met. More complex examples are treated in similar fashion.

POTENTIAL FOR FURTHER IMPROVEMENT

Further improvements to process conditions might be considered if the scale of the project had warranted it. The total operating time can be reduced by making the subsequent injection at a time before the elution of the last fraction collected equal to the time between injection and the appearance of the first-eluting component of the new run. This operation does not permit the removal of late-eluting impurities,

Table 7.3 — Composition of fractions of preparative-scale separation

Fraction	Amount A	Amount B	Purity A%	PurityB%	Recovery %
1	160	—	100	—	
2	1420	5	99.6		
3	2160	10	99.5		
4	1280	15	98.8	1.2	
5	630	40	94.0	6.0	
6	160	125	56	44	
7	90	270	25	75	
8	70	920	—	92.9	
9	35	970	—	96.5	
10	20	830	—	97.6	
11	10	470	—	97.9	
12	—	90	—	100	
13	—	90	—	100	
14	—	35	—	100	
15	—	20	—	100	
2–4	4860	30	99.4		81
2–5	5490	70	98.7		91.5
2–6	5650	195	96.7		94.0
7–13	225	3750		94.3	93.7
8–13	135	3480		96.3	87.0
8–15	135	3535		96.3	88.3

which will appear in fractions of subsequent runs. It is thus of greatest importance to define the problem carefully: if such impurities are unimportant in the use to be made of the product, or if they can be easily removed by a different operation, the saving in time and eluent can be considerable. This particular technique does not seem to be reported, but I have used it extensively when forcing the capacity of limited available equipment. The void volume elution time represents unproductive use of both stationary and mobile phase, as well as equipment and labour. Why make baseline when you can be making money? Be wary however, if using a UV detector, of early-eluting UV-invisible impurities — the void volume is often defined by the first UV-visible compound or baseline disturbance, and UV-invisible compounds, salts or high-boiling solvents or synthetic by-products, may be present in fairly large quantities, and may inhibit subsequent crystallization or increase toxicity. A step or continuous gradient to elute the second isomer more rapidly might also be considered, thus reducing the tailing and increasing the concentration of product, making it cheaper to recover. However, particularly with silica columns, the time taken to re-equilibrate the column will certainly more than offset the reduced separation time. In terms of ease of eluent preparation and recovery, total operation

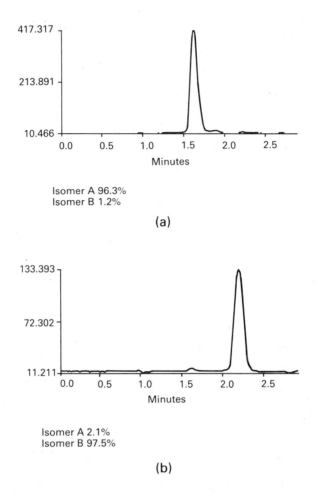

Fig.7.5 — Analysis of fractions 4 (a) and 11 (b) of preparative separation.

time as opposed to chromatography time, and reliability and ruggedness of operation there is much to be said for using isocratic procedures wherever possible in large-scale chromatography, as discussed below.

HARDWARE CONSIDERATIONS: THE DETECTOR

Determining the operating parameters for a large-scale liquid chromatographic separation, or purification, cannot be divorced entirely from a consideration of the hardware. If the project is a small or short-term one, it will almost certainly be performed on existing equipment, and the constraints of this will be built-in to the method development, with the possibility of examining whether addition of new

Ch. 7]   **Determining operating parameters in process chromatography**   117

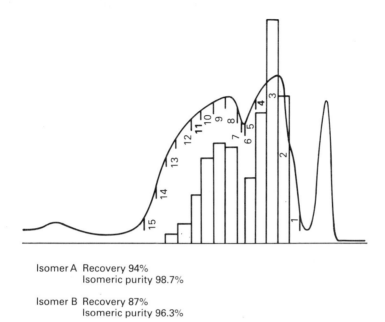

Isomer A Recovery 94%
          Isomeric purity 98.7%

Isomer B Recovery 87%
          Isomeric purity 96.3%

Fig. 7.6 — Summary of fractions combined from 10 g preparative separation, showing amount present.

concepts such as detectors or gradient equipment can be justified. For the definition of a full-scale manufacturing process, the specification of the equipment should develop in parallel with the method, and the method development should be directed not only to meeting the product requirements of the operation, but also to meeting the production considerations. The choice of detector type is one of the questions which influences, or is influenced by, the chromatographic method. The most obvious example is the choice between refractive index detector and ultraviolet detector. The basic characteristics are clearly differentiated: refractometers are less sensitive to product, but more sensitive to eluent composition than ultraviolet detectors — with the proviso, often forgotten, that the product has UV chromophore and the eluent components do not. So, for example, developing a method for a preparative system with only a UV detector will preclude the use of certain solvents as eluent components, such as acetone, and may enforce the use of a very-high-quality solvent to keep the eluent background absorbance to acceptable levels. Conversely, if only a refractometer is available, this clearly precludes the development of a gradient method.

It should be borne in mind that the essential function of a detector is different in analytical chromatography than in preparative chromatography: for analysis, the need is to determine *how much* of one or more components is present, and the principal detector requirements are sensitivity and linearity of response, coupled with as much selectivity for the analytes of interest as possible. For preparative work the essential function is to establish, and in some cases to document, *where* in the process one is, to permit collecting the desired product with the expected purity. This will usually, but not necessarily, involve detecting, or seeing, the component of interest. In a high-efficiency separation of closely related compounds, it is difficult to imagine monitoring the operation without directly visualizing the components, but in other cases, for example ion-exchange purifications, it may be sufficient to monitor the change of an eluent parameter such as ion strength or pH, having established during the method development that the required component elutes at a given stage. Of course, direct visualization always provides a useful verification of the presence of the product and its concentration, confirming that both the input feedstock and the separation process are within the expected limits. Thus the UV detector retains its popularity, especially for short-term preparative operations where the extensive method-stabilization necessary for full process operations may not have been established. The question of fixed wavelength or variable wavelength will then arise, and there are some considerations which are pertinent to their use at the large scale which should be examined.

Fixed- or variable-wavelength UV detectors
These considerations may be instrument-related or product-related. One disadvantage of the UV detector from a preparative point of view is its high sensitivity. This of course explains why other analytical detectors, such as fluorescence or electrochemical detectors, are not relevant to preparative chromatography. The variable-wavelength detector has the facility of allowing, for analysis, the selection of either the wavelength of maximum sensitivity or the compromise of 'de-tuning' to give less sensitivity for preparative use. With the advent of preparative flow cells of short pathlength and low flow resistance, the variable-wavelength UV detector became a much more useful preparative detector. However, this facility should be used with caution, particularly if the absorbance, or the absorbance slope, are to be used for triggering fraction collection.

The absorbance maximum is usually thought of as the wavelength of greatest absorbance — which, of course, it is — but it is also the wavelength of zero rate of change of absorbance, which makes it the wavelength least susceptible to variations in absorbance intensity owing to instrumental or operational variations. Fig. 7.7 shows the UV spectrum of a compound chosen at random, and the effect on the absorbance measured of a wavelength error. The wavelength of half-intensity $\lambda_{1/2}$ is also the wavelength of greatest rate of change of absorbance with wavelength, and is thus the most sensitive to error. In the example shown in Fig. 7.7, a 13 nm wavelength change is necessary to change the absorbance by 1% at the absorbance maximum, but the same change at the half-height produces a change of 25% in the absorbance! Of course, 13 nm is much more than the error of any detector on the market, but there are other causes of error than the specification of the equipment: regular maintenance and calibration is necessary to maintain the manufacturer's specification. There is also the possibility of accidental mis-setting of the wavelength, of forgetting to verify the wavelength if other use is made of the detector between batches, or of systematic differences between different users. Even the possibility of deliberate mis-setting must be considered if the process is a manufacturing operation

Ch. 7]   **Determining operating parameters in process chromatography**   119

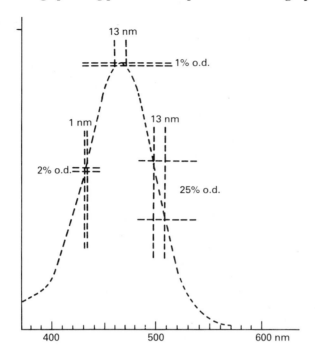

Fig. 7.7 — UV spectrum showing effect of wavelength errror on absorbance measurement at peak maximum and half-height.

in the production of of a registered compound. The compound shown in Fig. 7.7 does not have a particularly narrow UV spectrum, and more extreme cases are easily envisaged.

There may also be sources of change in wavelength other than instrument-setting errors. The absorbance maximum is dependent on solvent composition and pH, and pH is in its turn dependent on eluent composition, temperature and, of course, accuracy in eluent preparation or gradient composition. The effect of pH change on the spectrum of the compound in Fig. 7.7 is shown in Fig. 7.8. Such a change of 0.1 pH units may be caused by a 4°C change in temperature of an organic buffer such as a TRIS salt [12].

These changes of a few per cent in the absorbance of an eluting compound will not in themselves usually be serious if the separation is being followed by an operator, and the fractions collected by hand, although some separations are extremely sensitive to the position of fraction cutting; for example, if an impurity appears on the leading edge of the elution peak of the product. The position is more serious if the absorbance level or slope is being used to trigger automatic operation. The effect of a 2% change in absorbance on the relative proportions of fractions is shown in Fig. 7.9: the effect on the product purity will of course vary from case to case.

These considerations are not presented to suggest that 'de-tuning' should not be used: it is indeed preferable to have a detector output somewhat sensitive to changes

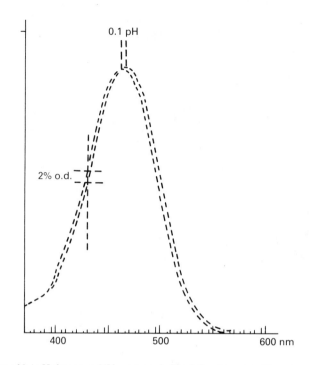

Fig. 7.8 — Effect of 0.1 pH change on UV spectrum in Fig. 7.7, extrapolated from spectra at pH 2 and pH 10.

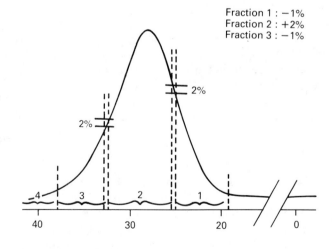

Fig. 7.9 — Effect of absorbance error on fraction composition.

in equipment condition or operating parameters, than to have an out-of-range signal because the detector is too sensitive. Rather the discussion is intended to alert the user to the potential dangers. Certainly no preparative separation, not to mention any manufacturing process, should be atempted without knowledge of the UV spectrum of the components and the sensitivity of the absorbance to wavelength errors at the chosen wavelength. Likewise, it is not sufficient to adopt the wavelength of an analytical method without considering the form of the UV spectrum; most analytical detectors operate at 254 nm or 280 nm, regardless of the UV spectrum, and regardless of the possibilities offered by the apparatus — indeed it may be said that a variable-wavelength UV detector is one which offers the user the freedom to choose to operate at 254/280 nm!

Examination of the UV spectrum may suggest a subsidiary wavelength maximum of lower intensity, which usually occurs at longer wavelengths. In addition to the advantages of lower concentration sensitivity resulting from lower extinction coefficient, and lower sensitivity to wavelength error, operating at longer wavelengths has the undoubted advantage in large-scale operations of lower backgrouund absorbance, which leads directly to lower costs of eluents.

Once the question of wavelength has been resolved, the choice between fixed wavelength and variable wavelength can be resolved in favour of fixed wavelength, if a suitable wavelength is available, because of the reduced susceptibility to error; if not, or if no alternative is available, then a variable-wavelength detector is favoured. In the case of chromatography as a manufacturing process for pharmaceuticals a fixed-wavelength detector presents distinct advantages from the point of documentation and security of operation, or the possibility of locking the monochromator at the chosen wavelength can be considered, otherwise the manufacturing protocol should incorporate a verification of the setting and calibration of the wavelength.

Eluent regimes: isocratic or gradient
Another factor where hardware considerations can affect the method, and should therefore be taken into consideration from the outset, is the eluent regime: isocratic, step-gradient or continuous gradient. There is no doubt that from an operational viewpoint an isocratic separation is best, and if that single eluent can be a single solvent, or azeotropic mixture, so much the better. However unlikely that may be, the possibility should not be forgotten, especially for straight- or reverse-phase separations using distillable solvents and not dependent on pH or ionic strength! However, there are inherent disadvantages to isocratic methods: peak broadening due to volume overload, tailing peaks, retention of impurities on the column, excessively long separations and potential for error, human or mechanical, in the preparation of the eluent. On the positive side are column stability, lack of column equilibration time (assuming no wash step at the end of each run), possibility of recycle, simplified hardware, greater choice of detector (refractive index or density) leading to greater freedom of choice of eluent component identity and quality, stable baseline and lack of degassing or mixing effects. The latter depend to some extent on whether the eluent is premixed or is created continuously during chromatography — a sort of continuous gradient of zero slope. Perhaps most importantly, a potential source of batch rejection, non-conformity of the gradient profile with the protocol,

and the data storage and documentation associated with gradient operation, is eliminated. This raises some of the same questions as the consideration of gradient operations, but it will be well to consider them here, and to elaborate to the continuous gradient case later. Step gradients can be regarded as a series of isocratic separations, with the discussion of pre-mixed or generated eluents a multiple of the same concerns as for an isocratic operation. However, there are some problems particular to the step gradient, notably problems of mixing and degassing as the eluent boundary progresses down the column, and baseline disturbance, even to 'gradient-compatible' detectors, which are not necessarily insensitive to refractive index changes.

Pre-mixing or on-line mixing of isocratic eluents
The question of pre-mixing as opposed to continuous generation is an important one at the process scale, more so than at the small scale of analytical separations, where problems of solvent handling, accuracy of measurement and efficiency of mixing are minor, and trained technicians using accurate volumetric glassware and analytical balances can achieve reproducible accurate results and can assure complete mixing and proper degassing, probably while supervising the operation of the chromatograph itself. It should not be assumed that the same operation is as easy when carried out with volumes measured in cubic metres!

In the manufacturing environment the two possibilities present conflicting problems and disadvantages, and as always when two opposing interests have to be considered, it is impossible to give a general conclusion, only to outline the consequences of each procedure and facilitate the final decision in any specific case. From the standpoint of the chromatograph specification, premixing is undoubtedly the method of choice; the chromatograph will be simpler and cheaper to buy, to install, to use and to maintain if all solutions — feedstock, eluent or eluents and wash solutions — are provided pre-mixed, degassed and equilibrated in temperature. In some cases, this may even obviate the need for explosion-proof equipment.

The manufacturing environment includes more than just the chromatograph however. The process engineer will certainly prefer a 'black box' which has only to be fed with solvents and solutions directly available from tank or drum, leaving to the machine the task of producing the exact compositions required for the operation. Such an approach will reduce the potential for machine or human error in the preparation of eluents. What then are the disadvantages of leaving to a gradient system the task of producing required eluents from bulk solvents? The reasons may be grouped under two headings: those related to mixing solutions *per se* and those related to the chromatographic equipment.

Mixing and degassing problems
Solvent mixing phenomena will exist whether the mixing device is a two-pump high-pressure mixer or a proportioning-valve single-pump low-pressure mixer, but the consequences will differ. The effects that cause concern are incomplete mixing, degassing and temperature changes due to heat of mixing. Because it occurs by definition after the pumps, in a volume of tubing which should be as small as possible, particularly if the feedstock is also to be applied through one of the pumps, mixing

would be expected to be a more serious problem in high-pressure mixing systems, and this would be exaggerated by the fact that dead volumes, as a proportion of column void volume, are already much smaller in process chromatographs than analytical chromatographs, which reduces band-spreading during injections for example, but also reduces available mixing volume. This is to some extent counteracted by the fact that turbulent flow is more readily achieved in large systems. The disadvantage of reduced time available for mixing in high-pressure mixing systems is balanced by the reduced significannce of degassing by the mixed solvents. The main problem with eluents degassing after mixing is the effect on the pump heads, and thus on the reliability of the separation. The effect of gas (air) in membrane pump heads is simple: they stop working. The incorporation of a flow meter will provide a warning, but will not prevent the effect, and if it is a feed-back control it will tend to just increase the pump rate, to no avail.

The effect of heat of mixing is one that will probably pass unnoticed during small-scale experiments, either analytical or as part of the scaling up process. This is because analytical chromatographs with $\frac{1}{16}''$ tubing have a large surface area of tubing per unit volume and react rapidly to changes in ambient temperature. A sudden ray of blessed sunlight in spring has a beneficial effect on the morale of the laboratory technician, but a less advantageous one on the equipment, particularly refractometers. On the contrary, changes in inlet eluent temperature will be rapidly equilibriated (or there would be no need for column ovens!). For process systems the large mass of solvent, and the low specific surface area will result in lower sensitivity to ambient changes, but greater retention of heat of mixing. For example, for a chromatograph with a 10 cm diameter column and 6 mm o.d. tubing, compared with an analytical chromatograph, a 4 mm column and $\frac{1}{16}''$ o.d., 0.009'' i.d. tubing, there is over 600 times the amount of heat to be dissipated under equivalent operating conditions, but the surface area of the tubing per unit length is not even four times greater! Actual tubing lengths are probably shorter for the larger apparatus, depending on the type of pulse-damper in the analytical pump. The heat of mixing can be negative (acetonitrile/water) or positive (alcohols/water), and can be considerable: I have measured temperatures in excess of 30°C at the head of the column when making methanol/water mixes. This can have an effect on either the separation itself or on the equipment, such as condensation on optical surfaces and back pressure due to the viscosity change. As an indication of the type of effect on chromatography that might be expected, I have found that changing the operating temperature of a fatty-acid separation from 20°C to 5°C required a change from 95% methanol/water to pure methanol. A 10°C change in temperature will change the pH of a 0.05 M TRIS buffer by about 0.3 pH units [12], with possible consequences on α values, or on absorbance values and slopes, and so on fraction-cutting points. An example of the kind of change that can occur is shown in Fig. 7.10, showing the need, when making preparative separations, of investigating the conditions around those at which the separation is planned, to chart the susceptibility to change. If these observations were transferred to preparative separation, the α-value of 1.30 found between homovanillic acid and benzoic acid at pH 4.0 crashes to 1.07 at pH 3.5. The steep part of the k' vs. pH curve is not always fraught with danger: in the same example the curves for salicyclic acid and homovanillic acid are almost parallel

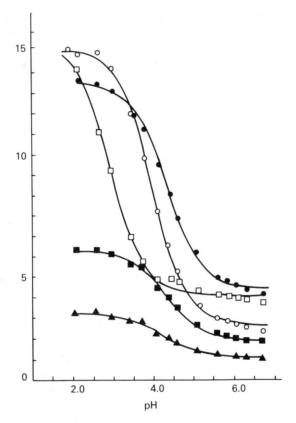

Fig. 7.10 — Capacity factors of weak acids as a function of pH. Reproduced from *Analytical Chemistry* **49**, 150 (1977).

although steep over a large range, so the α-value would be much less sensitive to eluent variation.

HARDWARE CONSIDERATIONS

The hardware-related concerns are directed at the proportioning valves in single-pump systems and at the flow-rate range and accuracy in two-pump systems. Let us look at the proportioning valve system first.

The creation of a mixture of desired composition by mixing two (or more) inlet streams by proportioning valves is based on the principle that the volume of solvent admitted to the inlet side of the pump is proportional to the time the valve is open. This assumes that the viscosity of the two liquids is equal and that neither is degassing, that the valve performance is constant, providing as close to a square-wave slug of solvent as possible, and that the pump has a steady-flow inlet profile. Despite the manifest absurdity of these conditions, manufacturers provide systems that produce very accurate gradient profiles. However, it is well to consider the limitations inherent in the method, in order to avoid making unreasonable demands, or of developing a separation method which is too critically dependent on performance. The cycle time of the valves of a low-pressure gradient system is typically a second or two. The only type of valve capable of offering this rate of operation is a solenoid valve. At a composition of 1%/99%, the valve controlling the 1% inlet will be open for 10 or 20 milliseconds. This is of the same order as the time taken to move the solenoid from open to closed, or vice versa. It is clear that this is making great demands on the reproducibility of the valve operation and the controlling signal generator. A further consideration is that while solenoid valves are highly reliable components, and serious manufacturers will test them under operating conditions for many millions of cycles, they do contain both moving parts and seals, and are therefore subject to maintenance requirements. Generally therefore, some extreme part of the composition profile, say 0→10% (and 90→100%), is less accurate and reproducible, and should be avoided if possible, or at least the sensitivity of the method to compositional variation determined and if possible minimized. Unfortunately, it is eluent compositions such as 5% water in methanol (fatty acid separations) or 5% ethyl acetate in trichlorotrifluoroethane as above, that are most sensitive to changes in concentration of the minor component. Hence the interest of considering during the scale-up stage these factors of how to accomplish the process operation, which might otherwise be thought irrelevant to method development.

In the case of high-pressure mixing using two pumps, the mechanical causes are different, but the operational consequences surprisingly similar. A good, frequency-controlled, analytical liquid chromatography pump can be set to deliver from 1% to 100% of its rated flow rate, typically 0.1–10 ml/minute (or better; detailed specifications are not at stake here), and it will reliably deliver much lower flow rates when controlled by the frequency signal of a gradient generator. Therefore, as far as volume composition is concerned, this configuration is regarded as extremely accurate, if rather expensive. It would be natural to believe that a gradient (or an isocratic composition) developed with such a system could be simply scaled-up to larger volumes. Unfortunately this is not so. The pumps used for process-scale chromatography are usually membrane pumps. These are volumetric displacement pumps whose accuracy is usually about 1%, but this is valid only over the range of about 10→100% of rated flow rate. Below this flow rate the accuracy falls off considerably, and to a degree depending on the solvent compressibility and gas content, all flow finally stopping at some point about 5% of the rated maximum. Thus a composition lower than about 10% cannot be expected to give an accuracy anywhere near 1%, and lower than 5% is probably impossible. Two mechanisms are used to control flow rate, varying either stroke length or motor speed. The above considerations are valid for the stroke adjustment. For reasons of power output and heat dissipation frequency control of motor speed is similarly limited. Clearly the two mechanisms can be combined to give a greater range, but this will make for a more expensive and more complex chromatograph. A further disadvantage of the two-pump system is that for the full range of composition and flow rate, the installed pump capacity must be double the nominal maximum flow rate. Of course one pump may be of lower capacity, say 20% or 50% of the other, extending the accurate range down to about 2%, but limiting the maximum composition correspondingly.

For separations intended only for the preparative scale (flow rates to a few hundred ml/min), the performance of the pumps can be regarded as equal to analytical pumps.

Thus, whichever mixing method is used, compositions of less than 10% of one component should be generated from premixed solvents, such as 20/80 or 50/50 dilutions or from an accurate eluent composition requiring no further blending. There is some advantage of using standard premixes in the manufacturing environment (indeed if the scale is large enough they may be available from the supplier), rather than exact fractional compositions such as 13.5%, in order to reduce the liability to operator error. However, it might also be argued that if you are going to do some premixing you might as well do it all, and benefit from a simpler and cheaper chromatograph!

The advantages and disadvantages of off-line and on-line preparation of isocratic eluents are compared in Table 7.4.

Table 7.4 — Comparison of off-line and on-line isocratic eluent generation

Advantages	Disadvantages
Off line generation	
Accuracy	Depends on operator reliability
Constant composition throughout run and run-to-run.	
Possible to monitor eluent batch by off-line analysiss (HPLC or other)	May vary batch-to-batch
Simpler chromatograph: cheaper, simple to operate, fewer parameters to monitor, less data to record	Requires equipped mixing-room, possibly Ex
Better mechanical reliability: no degassing in pump-heads, no valve maintenance	Handling/transport of solvents
No heat of mixing	Attention to temperature of eluent supplied!
On-line generation	
No operator error in eluent prep	Composition may fluctuate during run or run-to-run
Composition can be easily changed during run or between runs	More complex chromatograph: more expensive, more maintenance, more complex operation, more data monitoring and storage.

In summary, premixing of eluents for isocratic separations is preferable to on-line generation, provided always that the circumstances of the particular manufacturing operation permit safe and reliable off-line pre-mixing.

When developing a method for scaling up to a large-scale process, especially if it is to be a full-scale manufacturing operation, the constraints likely to apply at the large scale should be considered when the type of method is being decided. If, for example, the manufacturing facilities are such that off-line pre-mixing of eluent for an isocratic separation is impossible or unlikely, then the chromatographic equipment necessary for on-line eluent generation will not be very different from that needed for gradient separation, and there are therefore fewer advantages to restricting method development to isocratic conditions. The relative advantages of isocratic and gradient separation are summarized in Table 7.5.

GRADIENT ELUTION REGIMES

In general terms, if a gradient method is used, there are few restrictions on scaling it up to full manufacturing scale, except that three- or four-component gradients should be avoided, unless it is known that the large-scale equipment will have the necessary capability. If it is possible to develop a gradient method in which the correct operation of the gradient can be verified and documented during the separation itself, rather than before or after, this will be an advantage for documentation of the correct operation of the manufacturing process. An example would be monitoring of the pH or conductivity of the eluent, while the product elution is followed by UV absorbance. A further possibility is the monitoring of the density profile between pump and column, which would not depend on ionized species nor interfere with UV measurement, but I am not aware of this actually having been used.

When putting a gradient method into operation at the large scale, two conditions should be avoided: low percentage mixtures (say, 0–10% or 90–100%), and very shallow gradients. The concerns about low-percentage mixes are exactly the same as in the case of on-line isocratic mixing: the accuracy of pump-delivery rate, or proportioning valve opening and closing patterns, in the lowest part of the operating range. The solution is also the same: use suitable pre-mixed eluent components to increase the percentage necessary. The advantages and disadvantages of using pre-mixed components or pure solvents are summarized in Table 7.6.

Shallow gradients

With shallow gradients, for example 25% to 28% acetonitrile in aqueous buffer the problem in using pure aqueous buffer and 100% acetonitrile is that only a limited part of the operating range is being used, and clearly the absolute error will be magnified. Degassing, cooling and mixing problems will also be at their most acute, and there is the possibility or local precipitation of buffer salts [13]. Clearly, to use initial pre-mixed components of 25% and 28% and a 0→100% gradient would

Table 7.5 — Comparison of isocratic and gradient eluent methods

Isocratic or step-gradient	Continuous gradient
Operation	
Simple and reproducible	Loading of dilute feedstock
Less documentation	Better peak shape, less tailing
Off-line confirmation of eluent composition — % components, pH, etc.	More concentrated product
No equilibration between runs, possible to overlap runs in void volume	Accommodates wider range of retention in shorter time
Recycle operation possible	Easier removal of retained impurities from column
	pH or ion-strength gradients needed for ion-exchange methods
Peak broadening due to volume overload	Time advantage often lost in equilibration step
Tailing peaks — product dilution	At least two eluents to prepare
Large k' range — slow separation	Eluent components may degass or heat/cool on mixing
Equipment	
Simpler, cheaper, less maintenance	More complex equipment is more expensive, needs more maintenance and monitoring during operation
Wider choice of detectors	
Easier solvent recovery	
If using gradient equipment: simpler operation, less maintenance fewer components to verify	gradient parameters, lag time, dead volumes less reproducible with different installations

provide absolute accuracy (within the limitations of the pre-mixing operation) at the initial and final points, but the accuracy of the early and late stages would suffer from operating over the less accurate 0→10% and 90→100% ranges, and therefore a gradient of say 10→90% or 20→80% using one initial eluent somewhere less than 25% and the other somewhere more than 28% would give the best solution. How to calculate the required reservoir concentrations or initial and final gradient set points?

Table 7.6 — Comparison of gradient eluent generation methods

Advantages	Disadvantages
Using off-line pre-mixed components	
Improved accuracy in low % range	All disadvantages of off-line isocratic eluent prep (Table 7.4)
Improved accuracy in small range	
Less degassing in pump-heads	
Less heat of mixing (hydration already achieved partially)	
Possibly better mixing	
On-line generation from pure solvents	
No operator error in eluent prep	Reduced accuracy and reproducibility in small range or low % range
Composition immediately clear during operation. Gradient condition changes easier to understand	Degassing and heat of mixing
	Possible mixing problems.

CALCULATION OF RESERVOIR CONCENTRATIONS FOR ELUENT PRE-MIXES

Let us take the case of a gradient made from a starting mixture of 90% of weak aqueous methanol, containing C_1% methanol and 10% of strong aqueous methanol containing C_2% methanol and going to a final mixture of 90% of the stronger aqueous methanol solution. The actual methanol concentration at the starting point will be:

$$C_{initial}\ (=C_i)=[(90\times C_1)+(10\times C_2)]/100 \quad (7.3)$$

and the final concentration will be:

$$C_{final}\ (=C_f)=[(10\times C_1)+(90\times C_2)]/100 \quad (7.4)$$

These equations can be put in the general form, for a gradient form P_1% to P_2% of the stronger solution:

$$C_i = [(100-P_1)C_1 + P_1C_2]/100 \tag{7.5}$$

$$C_f = [(100-P_2)C_1 + P_2C_2]/100 \tag{7.6}$$

The usual situation however is that the desired initial and final concentrations of the eluent are known, and the necessary reservoir concentrations to give a 10–90% gradient are required. Eqs (7.3) and (7.4) can be written to express the required reservoir concentrations C_1 and C_2:

$$C_1 = [(90 \times C_i - 10 \times C_f)]/80 \tag{7.7}$$

$$C_2 = [(90 \times C_f) - (10 \times C_i)]/80 \tag{7.8}$$

These can be written in the general form for the case of gradients symmetrical about 50%, that is where $P_2 = 100 - P_1$:

$$C_1 = P_2C_i - P_1C_f)/(P_2 - P_1) \tag{7.9}$$

$$C_2 = C_2C_f - P_1C_i)/(P_2 - P_1) \tag{7.10}$$

Where P_2 is not $100-P_1$, it is simpler to solve Eqs (7.5) and (7.6) for C_1 and C_2 directly.

Thus, for instance, we can determine the reservoir strengths necessary to generate a gradient of 25→28% acetonitrile using the optimum 10→90% gradient capability of the equipment:

$$C_1 = [(90 \times 25) - (10 \times 28)]/(90-10) = 24.625\% \text{ acetonitrile and}$$

$$C_2 = [(90 \times 28) - (10 \times 25)]/(90-10) = 28.375\% \text{ acetonitrile.}$$

If premixing is to be done, certain limits will probably be placed on the reservoir concentrations, such as whole-number or round-number concentrations, or whole-number weight ratios, with a view to reducing operator error. In this case Eqs (7.5) and (7.6) can be written to express the necessary gradient set points P_1 and P_2 needed to generate the required gradient from any suitable starting reservoir concentrations:

$$P_1 = 100(C_i - C_1)/(C_2 - C_1) \tag{7.11}$$

$$P_2 = 100(C_f - C_1)/(C_2 - C_1) \tag{7.12}$$

For the above case, if convenient reservoir concentrations are 20% and 30% acetonitrile, the starting- and end-points become:

$P_1 = 100 \times (25-20)/(30-20) = 50\%$, and
$P_2 = 100 \times (28-20)/(30-20) = 80\%$.

Thus a gradient should be programmed from 50→80% and reservoirs of 20% and 30% acetonitrile prepared. Although these particular concentrations use only 30% of the total gradient range available, this is a considerable improvement on using only 3% of the range, and will increase accuracy and reproducibility accordingly.

CONCLUSION

The final conditions determine the success of the operation [14].

To the chemist considering scaling up a separation these questions of solvent properties, method ruggedness, detector type and gradient-generation hardware may seem irrelevant, and the means by which the process engineer finally puts into operation the conditions defined, so long as it has been shown that the process does indeed scale-up correctly, not to be his concern. Consider however the effort to be expended on carrying out the scale-up procedure, the number of preparative runs, the associated analytical evaluation of the fractions, possibly subsequent operations carried out on the product of an intermediate-scale preparative operation to produce some hundreds of grams of investigational material, perhaps even some preliminary toxicological studies. Consider then the position of the production team that is forced to resolve the dilemma: 'we cannot use acetonitrile without changing our ventilation equipment and introducing a work-place air quality monitoring scheme, and we would like to use ethanol because we already have a holding tank and the necessary customs licences, but we do not know whether the purification would be satisfactory, and the team who did the scale-up have been assigned to another project!' Then I think that my view will be shared that the parameters of the end-process are as much of concern in the devising of a method *suitable for production* as are the mechanics of the process by which the larger scale is developed from the smaller.

REFERENCES

[1] P. D. McDonald and B. A. Bidlingmeyer, 'Strategies for successful preparative liquid chromatography', Chapter 1 in *Preparative liquid chromatography'*, *Journal of Chromatography Library* Vol. 38, Elsevier, Amsterdam 1987.
[2] M. Kelly, Letter to the Editor, *Pharmaceutical Technology International*, **3**(4), 11 (1991).
[3] M. Kelly, 'Measurement and control in process liquid chromatography', in *Measurement and control in bioprocessing*, Elsevier Science Publishers, London 1991.

[4] G. Carta, H. Massaldi, M. Gregory and D. Kirwan, 'Chromatography with permeable supports: theory and comparison with experiments', submitted for publication 1991.

[5] N. B. Afeyan, N. F. Gordon, I. Mazsarof, L. Varady, S. P. Fulton, Y. B. Yang and F. E. Regnier, 'Flow-through packing materials for high-performance liquid chromatography of biomolecules: perfusion chromatography', *J. Chromatography* **519**, 1 (1990).

[6] G. L. Hagnauer, 'Preparative size exclusion chromatography', Chapter 8 in *Preparative liquid chromatography*, see [1].

[7] American Conference of Government Hygienists, *Threshold limit values for chemical substances in the work environment* ISBN 0–936712–78–3 (1988).

[8] B. Hill, 'The manufacture and utilisation of electrical equipment for use in potentially explosive atmospheres', *Electrical and Mechanical Executive Engineer*, (May/June 1982).

[9] O. F. Pedersen and L. Buus, Poster presentation 'PREP 86', Paris (1986), 'Purification of LH–RH derivatives by HPLC on a process scale'.

[10] P. D. McDonald. Unpublished observations, cited in [1].

[11] J. K. Whitesell, 'Preparative liquid chromatography for the organic chemist' Chapter 5 in *Preparative liquid chromatography* see [1].

[12] *The ISCO Tables*, 8th Edition, Lincoln, Nebraska (1985).

[13] J. W. Dolan, 'Retention-time drift — a case study', *LCGC International* **4**(4), 20 (1991).

[14] 'and the final condition of that man is worse than the first', Matthew **12**, 43–45.

8

Design and cost implications in process-scale chromatography

Kevin Connelly
ICI Pharmaceuticals, Macclesfield, UK

INTRODUCTION

The increasing complexity of material developed in the speciality and fine chemicals businesses, coupled with the demands for higher purity, has provided the impetus for the development and exploitation of new technologies to facilitate both product recovery and purification.

Among these new technologies, process-scale chromatography has developed rapidly and is now accepted unit operation. Although still a techniquie of last resort in a production environment, it has been proven to have widespread applicability and is now routinely used by several large pharmaceutical companies in the manufacture of sales products.

The design of a process-chromatography system is dictated by the operating requirements of the column. However, in the total design the column represents only a small part of the system and a number of other factors must be considered to ensure overall operability and effectiveness. It is the purpose of this paper to consider these factors.

Fig. 8.1 shows a typical line diagram for a process chromatography system. It can be conveniently broken down to the following areas:

(1) Eluent preparation
(2) Sample preparation and loading
(3) Eluent pumping
(4) Column
(5) Product detection
(6) Fraction collection
(7) Product recovery

and each will be discussed in turn.

Eluent preparation

A characteristic of preparative chromatography systems is their relatively high consumption of solvent. The design of the eluent system is critical to the efficient operation of the process.

The system design should be capable of providing a reproductive eluent to the column at the required rate without the need to interrupt column operation. To achieve this, a number of factors which should be considered are listed in Table 8.1.

Column operating mode

The column may be operated in the classical sense whereby a number of injections of the material to be separated are loaded and eluted sequentially. In this case, eluent consumption is determined by the flow-rate cycle time and number of injections. In routine production, the unit may be expected to operate continuously over the working day.

Alternatively the column may be operated in an adsorptive mode, whereby the packing is saturated with product, which is then eluted, normally with a gradient. In this case, flow of eluent is intermittent.

Elution mode

The column may be operated in either an isocratic or gradient mode. In the former case, batches of eluent can be prepared and used. In the latter at least two vessels containing the components of the eluent will be required.

Column packing

The column can be operated with either reverse- or normal-phase packing materials. In the former case, a source of purified water is required. This may be conveniently provided by ion-exchange or reverse osmosis. It is not sufficient to use domestic water, which contains varying quantities of dissolved solids and inorganics that can rapidly foul the column packing.

Handling of solvents in significant quantities requires careful consideration. It may be necessary to charge from drums and the handling of drum traffic should be carefully thought out. Alternatively, it may be convenient to change from a bulk storage to the eluent vessels.

Eluent reproducibility

The performance and reproducibility of the chromatography column can be significantly affected by the eluent. In particular, temperature can have a marked effect and it may be necessary to consider temperature control of the feed vessel. If the unit is located in a heated building, ambient temperature is convenient, but it must be recognised that it can take a significant time to equilibrate, particularly if there are significant heats of mixing involved in the eluent make up.

The composition of the eluent can also have a dramatic effect on the chromatography. It is essential that batch-to-batch variations are minimized. Hence solvent

Ch. 8] Design and cost implications in process-scale chromatography 135

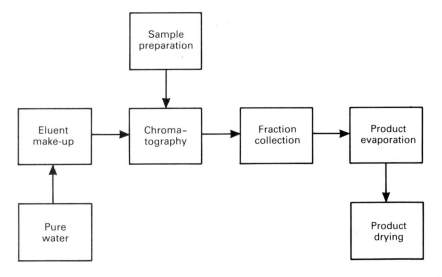

Fig. 8.1 — Process-chromatography system.

Table 8.1 — Factors to consider in eluent system design

Column operating mode
Elution mode
Column packing type
Eluent reproducibility
Eluent filtration

---

changes must be accurate and reproducible. This can be achieved by either charging by weight or volume. Since some variation is inevitable, it is preferable to minimize the number of batches of eluent required to achieve the processing of a batch of material.

Eluent filtration
It is essential to filter the eluent prior to the column to remove extraneous material and fine particles which would otherwise block the column. A cartridge filter (0.2 mm) of the pall type is suitable provided care is taken to select appropriate seal materials, which do not leach organics into the eluent.

Fig. 8.2 shows a typical eluent system for operation in an isocratic mode. Eluent is made up in one vessel and transferred by a pump via a polishing filter to the feed vessel. Solvent changes are by weight using residual vacuum. The transfer between vessels could be achieved by vacuum, but if this option is used the effect of reducing the suction pressure on the performance of the eluent feed pump should be considered, since it may cause NPSH problems and affect the run in progress.

If a gradient elution is to be operated, then two feed pumps will be required, together with a control system to ratio the feeds. An in-line mixer (static sulzer type) is used to ensure that the fluids are well mixed prior to entry into a column (Fig. 8.3).

Sample preparation

Sample preparation is a batch process. The quantities of material handled in processing a single batch may vary from a few grams to kilograms.

In general the sample is dissolved in either buffer or eluent and filtered prior to loading onto the column. The plant design should consider where this operation is to be carried out and how the sample is to be transferred to the column.

Eluent feed pump

The eluent feed pump is required to deliver solvent at a controlled rate and at pressures up to 130 bar. For this reason, variable speed gear pumps or variable displacement piston or diaphragm pumps are used. Gear pumps are generally reliable and have the advantage of producing a smooth discharge but are prone to damage if allowed to run dry. If piston or diaphragm pumps are installed, it is usual to use multiple heads and to smooth the flow with a pulsation damper. The pump should always be protected by a relief valve on the discharge side to prevent damage in the event of blockage. As with all high-pressure pumps, seal and diaphragm failure are the most likely problem areas. In critical applications, it is worth considering the provision of an installed spare.

Sample loading

On small-scale columns, it is usual to load the sample via a syringe. On a process scale, this is not a practical option and three alternative methods are available.

(1) Via as sample loop, which is filled using a small pump and then loaded on to the column by diverting the eluent flow through the loop. This method is useful for small loadings and has the advantage of giving reproducible volumes. However, it has the drawback that the loop is full of eluent after each injection. This must be either flushed clear with sample, which results in either dilution or loss of material, or blown clear with inert gas.

(2) Use of a separate pump to load the sample on to the column. In this case eluent flow is stopped, the sample loaded and the eluent flow restarted. Although this might seem to be a less than ideal method of loading it does not in practice cause any problems with the chromatography. The stop-flow method has a major advantage over the loop system in that the amount loaded can be easily adjusted.

Small, high-pressure dosing pumps (e.g. Prominent) are useful for this duty. It is important that sample loading can be carried out in a reasonable time,

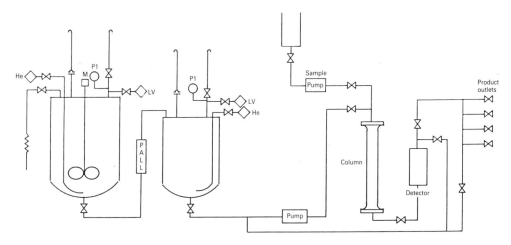

Fig. 8.2 — Eluent system for process chromatography in isocratic mode.

particularly when multiple injections are required and the pump should be sized commensurate with this duty.

(3) Loading via the eluent pump. By the use of a diverting valve on the eluent pump section, it is possible to load the sample using this pump. For multiple small injections, the method has the disadvantage that it may not be very reprodacible, and unless care is taken when designing the inlet pipework, mixing effects may dilute the sample. When larger volumes are loaded, particularly if the column is operated in an adsorptive mode, then this method is useful.

Column packing

There are a wide variety of column packings available. The selection of the most appropriate packing is dependent on the mixture to be separated and in general will have been specified during the chromatrography development. However, there are a number of factors peculiar to the process scale which should not be ignored.

Of paramount importance is the availability and reproducibility of the selected silica. Although there are large numbers of packings on the market, there are relatively few available in the quantities necessary to carry out preparative separations on a large scale. It is therefore essential that the development work is carried out using only those materials available in the amounts required. Close co-operation with a silica supplier is useful and can avoid subsequent problems. Packing stability is of major interest. In analytical systems, the dissolution of silica or bonded-phase is inconvenient over a period of time but is not a major problem. In process-scale work, chromatography may often be used as the final purification of the product. Silica or bonded-phase dissolving from the column, is certain to appear as impurity, although hopefully only in trace amounts.

There is currently very little published data in this area but it is undeniably one in which the regulatory authorities are likely to show increasing interest. Part of the development programme should therefore assess the loss of silica and bonded-phase

138   Design and cost implications in process-scale chromatography   [Ch. 8

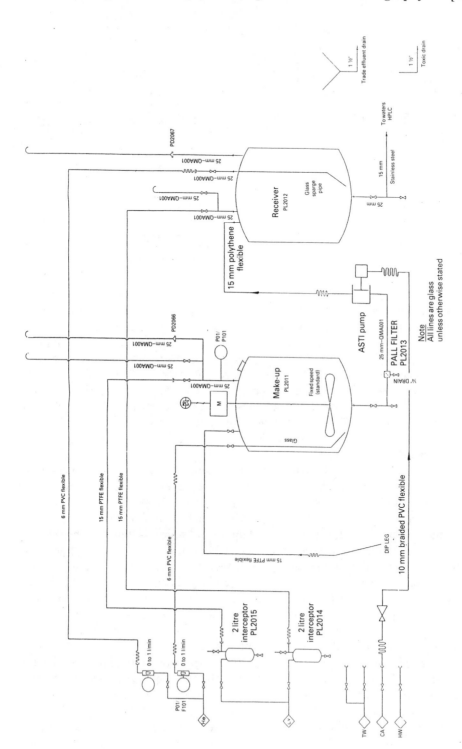

Fig. 8.3 — Eluent make-up.

Ch. 8]   **Design and cost implications in process-scale chromatography**   139

material. Methodology for carrying out these tests is not well developed and this is an area where silica suppliers can help. In testing for packing loss, it is important to consider all aspects of the operation. For example flushing the column periodically to clean and restore its performance, may have markedly different effects on packing loss.

Column life
Packing life can be a significant factor in the running costs of a process system, particularly when reverse-phase materials are used. Over a period of time, column performance will deteriorate owing to a number of factors.

(1) Build-up of particulates on the top of the column.
(2) Irreversible adsorption.
(3) Loss of efficiency caused by voiding arising from particle setting and attrition.
(4) Loss of bonded-phase from reverse-phase packings.

Column life is very dependent on the separation duty for which it is being used but some simple precautions can help to extend it.

(1) As described above, filtration of the sample and eluent can eliminate particulate build-up.
(2) If there are strongly adsorbed impurities in the sample, it is worth considering the use of a disposable guard column to extend packing life. It may also be worth pretreating the material by, e.g. flash chromatography.
(3) Column life can often be extended by periodic washing to remove strongly adsorbed species. The facility to back flush on the column is therefore a useful feature.

Column technology
The column technology offered by the principle manufacturers is summarized in Table 8.2. There is a choice between radial- or axially-compressed systems which are either self-packed columns or prepacked cartridges. There are advantages and disadvantages to each system and selection of the appropriate unit is dependent on the duty required. Some of the considerations are outlined below and in Table. 8.3.

(1) If the column selected is of the self-packed variety, then provision to handle packing and unpacking of silica must be made. The packing techniques vary but the handling of finally divided silica, with the inherent problems of dust control and static cannot be ignored. Similarly, unpacking the columns can result in silica for disposal which may be contaminated with biologically active materials.
(2) For multiproduct use, particularly in the pharmaceutical area, it may be unacceptable to use the packing for different products since it can be extremely difficult to demonstrate the absence of cross-contamination. In this environment, cartridges can offer advantages.

Table 8.2 — Pricipal manufacturers of columns for large-scale separation

Manufacturer	Packing technology	Particle size	Column size
Millipore Waters	Radial compression Prepacked cartridge	20–105 μm	100 m
Amicon	Axial compression Self-packed column Column packed in off unit packer	5–100 μm	Any can be provided
Prochrom	Axial compression Self-packed column Column packed in place	5–100 μm	50 mm 150 mm 300 mm 450 mm
Separations Technology	Radial/axial compression via Taper insert Self-packed column Column packed off unit	5–100 μm	75 mm 100 mm 150 mm 300 mm
Cedi	Axial compression Prepacked cartridge	5–100 μm 5–100 μm	25 mm 50 mm 100 mm
Du Pont	Prepacked columns	5–100 μm	25 mm 50 mm 100 mm

(3) It is now generally agreed that radial compression columns give only medium performance in terms of plate count. It has been demonstrated that axial-packed columns can produce plate counts greater than those obtained in analytical systems.

(4) Self-packing will not in general produce a good column every time. Experience within ICI indicates that one in three columns are acceptable although the reasons for variation of this are unclear. In general, cartridges are reproducible and can be supplied with a test certificate.

Detection

Detection of the product leaving the column can be achieved using UV absorption, refractive index, conductivity or, in the case of isomer separation, polarimetry. UV absorption and refractive index are the most versatile and commonly used of these techniques. Product detection can cause problems if care is not taken and some points to consider are as follows.

(1) Ensure that the detector is close to the product outlet. Most preparative separations do not give fully resolved peaks and rely on peak shaving to remove

Table 8.3 — Advantages and disadvantages of different systems

Manufacture	Advantages	Disadvantages
Millipore waters	1. Reproducible column 2. Easy replacement 3. Silica contained so easily handled if contaminated with toxic products	1. Limited range of packings and particle sizes 2. Limited range of column diameters 3. Fixed column length 4. Medium efficiency only
Amicon	1. Range of column sizes on single unit 2. Any material can be packed 3. High efficiency potentially	1. Separate column packer required 2. Axial compression cannot be adjusted without returning column to packer 3. Column reproducibility must be demonstrated
Prochrom	1. Easy to pack/unpack 2. Axial compression maintained 3. Any material can be packed 4. Range of bed heights 5. Range of column diameters available 6. Potential for analytical efficiency	1. Fixed column diameter on any given unit 2. Height constraints 3. Packing may not be reproducible
Separations Technology	1. Radial/axial compression maintained 2. Any material can be packed 3. Potential for high efficiency 4. Range of column sizes on single unit	1. Difficult to pack 2. Packing may not be reproducible 3. Bottom seal arrangement is weak
Cedi	1. Reproducible column 2. Easy replacement 3. Silica contained	1. Limited range of packings and particle sizes 2. Limited range of column sizes 3. Limited bed length
Du Pont	1. Reproducible columns 2. Easy replacement 3. Silica contained	1. Limited range of packings 2. Limited range of column sizes 3. Only column suppliers 4. Expensive, particularly if column-life short

poorly resolved impurities. If the distance between detector and collection point is excessive, then it may be difficult to correlate the cut-point with the fraction collected.
(2) Refractometers are very sensitive to small changes in ambient temperature. They are therefore difficult to use at high sensitivities and it may be necessary to enclose the unit in a constant-temperature box or at least prevent draughts.
(3) If a differential refractometer is used, then it is useful to incorporate a flush to the reference cell from the eluent feed line. It is then a simple matter to 're-zero' the unit to account for small variations in different batches of eluent.
(4) Split-flow detectors can cause problems. Split ratios of 10:1 or 15:1 are not uncommon but unless the flow through the detector can be recombined with the product stream then it may be necessary to recover and recycle the detector stream separately.

If the flows are recombined, then there is always a concern that the recombination is not exact and that peak mixing is occurring.
(5) Full-flow UV detectors with variable cell path length are versatile and can be operated at high sensitivity. This facility allows performance of the column to be assessed with small loadings and is a useful technique for monitoring column degradations.
(6) If an explosion-proof system is required it will be necessary to amount the detector electronics remotely or in a purged box.

Fraction collection

Fraction collection relies on the case of multiport valves to divert the product stream as required. The method of fraction collection is dependent on the ease of separation. For easy separations, it may be appropriate to collect in a reasonable-size vessel, combining the product out from a number of injections for further processing. It is more usual to collect small fractions by cutting across the peak. Each fraction is then analysed separately before combining for recovery or recycle. This inevitably leads to numerous small containers requiring interim storage space.

When the system is set-up initially, the determination of cut-points may require talking a large number of small fractions across the peak and analysing each to determine the separation. Any fraction collection system should include provision to do this.

Product recovery

The purified product is normally recovered in a number of dilute fractions, containing solvent or aqueous solvent mixtures which must then be removed. Frequently the products are thermally labile.

Evaporation under high vacuum is probably the most common technique for product recovery. This can be achieved on the small-to-medium scale with rotary evaporators (Fig. 8.4). On the larger scale wiped-film evaporators are used, but it may frequently be necessary to further concentrate to a solid using an RFE.

The techniques of ultra-filtration and reverse osmosis can also be used for preliminary concentration.

Ch. 8] Design and cost implications in process-scale chromatography 143

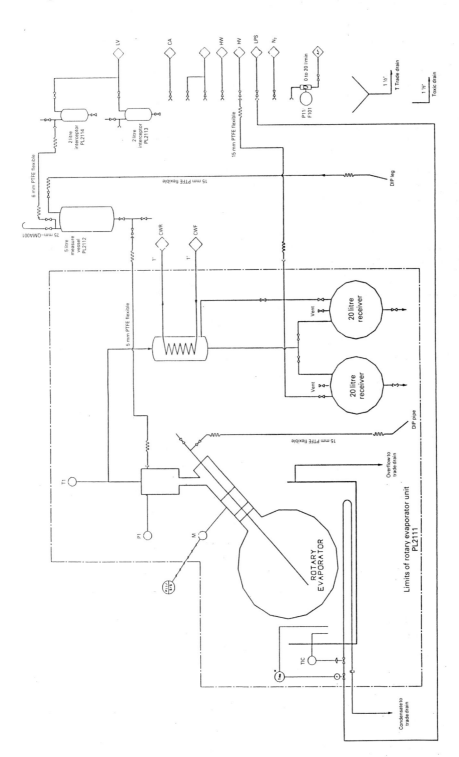

Fig. 8.4 — Solvent recovery.

It is important to provide sufficient evaporation capacity to handle the fractions produced by the chromatography system. This will in general necessitate multiple units, particularly if several different fractions are taken. Unless the units are dedicated, clean-out time between different fractions must be considered as an integral part of the cycle time.

Automation

The major suppliers offer automation packages of carying degrees of sophistication and cost. However, unless the separation is reproducible, they are likely to be of limited application. Without automation, chromatography is labour intensive so there is a significant benefit in developing reproducible feedstocks and operating conditions, particularly in dedicated plant.

Capital cost

It will have become apparent from the preceding discussion that the total capital cost is significantly greater than the basic system cost. As an example, I have taken the case of a Waters 250 kiloprep unit.

The unit is installed in the chemical production laboratory at Macclesfield Works. It is one of two units in the manufacture of a single product. The scale of operation is small (kg/annum) and the product has high added value. The plant is based on a large-scale laboratory principle and equipment is housed in 2 m fume cupboards. Operation is on a days only basis and personnel are skilled. The units are not automated. Table 8.4 gives a breakdown of the capital cost of the unit.

Table 8.4 — Capital cost: basis waters 250 kiloprep

	(£)
Eluent make-up	2800
(2×100 l glass vessels)	
Sample preparation	500
Chromatography unit	55000
(Feed and sample pumps, column, detector)	
Eluent collection	1000
Pure water system	2000
Rotary evaporators	25000
(2×20 l)	
Analytical HPLC	10000
1" System for scale-up	10000
Building	150000
(80 m^2 fume cupboards, services)	
Total	256300

Running costs

The running costs of a chromatrographic system are very dependent on the separation being performed and particularly in the case of reverse-phase silica, they can be very sensitive to column life.

An indication of the added cost of using chromatography on the plant described above is given in Table 8.5.

Table 8.5 — Operating costs

	Per gram processed (£)
Eluent	0.40
Steam	0.02
Electricity	0.08
Water	0.05
Column packing	5.20
Labour	8.33
Effluent disposal	0.61
Total	14.69

CONCLUSION

The overall design of a process-scale chromatography system requires careful consideration of all areas of operation to produce an efficient and integrated plant.

9

Factors of importance in the scale-up of ion-exchange process

Peter R. Levison
Whatman Ltd, Maidstone, UK

INTRODUCTION

The isolation of a pure target protein from a crude feedstock is routinely carried out in the bioprocessing industry following a defined series of unit processes. The number and sequence of unit processes is dependent on the composition of the feedstock, the degree of purity required for the target, the overall yield and general economic considerations. Low-pressure chromatography is an essential part of a downstream process and will often be carried out in several discrete steps, each relying on a different physical interaction between the components of the feedstock and the chromatographic medium. Dependent on the nature of the target material the chromatographic step referred to as process scale may be carried out in laboratory columns where only low levels of target may be present, e.g. mg quantities, or using large contactors where kilogram quantities of target are being purified. It is this latter area which is covered by this chapter.

Prior to carrying out any practical investigations, the protein chemist, typically in a research and development environment, is provided with a range of chromatographic techniques out of which he is challenged to develop an effective purification protocol. There are several established techniques available in low-pressure chromatography including those listed in Table 9.1.

In many instances the protein chemist will be offered a range of chromatographic media by several suppliers for a specific technique, with each product claimed to be superior to the rest. While all of the techniques listed in Table 9.1 are suitable for use at laboratory scale, scale-up may be problematic and impractical for a variety of reasons. These include media costs, availability and consistency of supply of bulk media, regulatory requirements, inclusion of additional unit processes to support the procedure, i.e. filtration, dialysis, etc. Of the chromatographic techniques listed, those typically scaled-up are ion exchange, hydrophobic interaction and size exclusion with the former being the most widely used.

Table 9.1 — Techniques available in low-pressure chromatography

(1) Salt precipitation
(2) Ion exchange
(3) Size exclusion
(4) Hydrophobic interaction
(5) Thiophilic interaction
(6) Affinity-chelation, biospecific, etc.
(7) Chiral

PRINCIPLES OF ION EXCHANGE

Ion exchange is a type of adsorptioon chromatography in which the adsorption of a charged solute to a stationary phase bearing the opposite charge takes place. Desorption is simply effected by exchanging the bound solute molecule with a counter-ion of similar charge. Ion-exchange stationary phases bear either a positive or negative charge for adsorption of either anions or cations, respectively. The selection of either anion- or cation-exchange media is entirely dependent on the nature of the solute ions to be separated although in protein separation anion-exchange tends to be the more widely used technique. In an aqueous environment proteins can be regarded as polyions and have an overall electrical charge dependent primarily on the secondary structure of the protein and post-translational factors such as glycosylation [1].

The isoelectric point, pI, of a protein is the pH at which it bears no net charge and this is dependent on structural aspects of the molecule. For pH>pI the protein bears an overall negative charge and thus binds to an anion exchanger and for pH<pI the protein bears an overall positive charge and binds to a cation exchanger. Since ion exchange is an adsorptive process, isocartic elution is not recommended and selective desorption is simply carried out by increasing the ionic strength and/or adjustment of the pH closer to the pI. Low-pressure ion-exchange media are available in bulk, and for application in the bioprocessing industry, are traditionally based on polysaccharide supports including cellulose, agarose and dextran [2,3].

Considerations for scale-up

In scaling up an ion-exchange process there will typically be three phases of development. Firstly, the chromatographic conditions will be developed at the analytical level and at laboratory scale. This separation would then have to be demonstrated to scale-up effectively. Secondly, the capacity of the ion-exchange medium would be determined under the predetermined adsorption conditions. Thirdly, the process would be scaled-up using the most efficient means of handling the media, i.e. batch or column. These initial stages would be developed at the research laboratory bench and several key factors need to be considered at this time.

These include:

Media availability
If the ultimate unit process is likely to require bulk quantities of ion-exchange media, e.g. tonnage, then factors including cost, continuity of supply, consistency of supply, etc., need to be investigated.

Quality of the media
In order to maintain consistent chromatography and product purity, media quality must be maintained. While the user of the ion exchanger will often carry out in-house quality assurance of the media, factors such as batch size, supplier's quality control test data and external quality accreditation of the media supplier, e.g. BS5750 (ISO 9001), should be investigated.

Regeneration and sanitization of media
In purification processes where re-use of the media and/or sterility are key factors then protocols for media regeneration and sterilization and depyrogenation should be identified at an early stage. The chemical compatibility and physical stability of the ion exchanger to these regimes needs to be confirmed in order to ensure that the ion exchanger is not destroyed during such procedures.

Provided that the ultimate process requirements can be met taking into account the factors described above, then the ion-exchange process should be appropriate for scale-up.

Throughput
The key to economic success in a process-scale ion-exchange step is throughput. This may be defined as the amount of product isolated at the required purity per unit time, i.e. kg/day. Throughput is influenced by media selection, capacity of the medium, adsorption/desorption kinetics of the medium and process time, i.e. flow rate.

Media selection
In an ion-exchange process the nature of the target protein and the pH and ionic composition of the mobile phase will influence selection of the chromatographic medium. Some of the characteristics of several Whatman anion-exchange celluloses are listed in Table 9.2. The small-ion capacity quoted in milliequivalents per dry gram is an empirical measure of the level of substitution of charged groups on the surface of the media support matrix. For a monovalent anion, e.g. Cl^-, the small-ion capacity would therefore be millimoles per dry gram. The data in Table 9.2 indicates that DE52 has approximately five-fold increased levels of derivatization over DE51, and DE53 has twice the level of substitution of DE52. In separations where the target molecule has a very strong negative charge, e.g. DNA, then DE51 may be appropriate. On the other hand where the target bears a very weak negative charge then DE53 may be more appropriate, but this may necessitate a pretreatment with DE51 or DE52 to remove the more strongly charged contaminants which would bind to DE53 preferentially to the target. The protein capacities of DE51, DE52 and DE53 for soybean trypsin inhibitor, bovine serum albumin and porcine thyroglobulin are summarized in Table 9.3. These data demonstrated that as small-ion capacity

Table 9.2 — Properties of Whatman anion-exchange celluloses

Grade	Small-ion capacity (mequiv./dry g)	Tertiary amine content (%)	Quaternary amine content (%)
CDR	0.30	100	0
DE51	0.22	100	0
DE52	1.00	89	11
DE53	2.00	48	52
QA52	1.10	0	100

Table 9.3 — Protein capacities of Whatman anion-exchange celluloses following batch adsorption using 0.01 M sodium phosphate buffer, (pH 7.5)

Exchanger	Protein capacity (mg/ml swollen medium)		
	Soybean trypsin inhibitor	Bovine serum albumin	Porcine thyroglobulin
DE51	21	8.4	4.0
DE52	138	86	6.6
DE53	155	91	6.9
QA52	168	107	8.9

increases, protein capacity does not increase proportionally. This may be expected since proteins are not small ions and steric factors, etc., come to bear. The protein capacity for DE53 is similar to DE52 but since the surface charge density is increased then the binding strength between the protein and DE53 would be increased compared with DE52. This increased binding strength may be exploited where two proteins of similar charge are to be separated. The more acidic protein would be retained longer on DE53 during a salt gradient elution than it would on DE52 owing to the stronger interaction and this would consequently give better chomatographic resolution. The data in Table 9.2 show increased levels of quaternary amine content along the series DE51, DE52, DE53 and QA52. The pK_a value for a diethylaminoethyl tertiary amine group is ~9.5 and if a separation is to be carried out at >pH 8.5 then a quaternary amine ion-exchanger would be recommended in order to exploit the ionic properties of the adsorbent.

Capacity of the medium
The capacity of the ion-exchange medium under the defined operating conditions directly influences throughput. Unlike the small-ion capacity, protein capacity is not

an empirical value and is dependent not only on the degree of substitution of functional groups but the molecular weight of the protein and the mobile-phase conditions. This is exemplified in Table 9.3 where the capacity of DE51, DE52, DE53 and QSA 52 was determined for soybean trypsin inhibitor (MW 20 100) bovine serum albumin (MW 67 000) and porcine thyroglobulin (MW 670 000) at pH 7.5. The data indicate that protein capacity reduces as molecular weight increases. At higher pH values, the protein capacity of the tertiary amine exchangers (DE51 and DE52) tends to reduce as the DEAE groups begin to deionize, effects not apparent for the quaternized products DE53 and QA52.

The mobile-phase conditions directly influence the capacity of the ion exchanger for the target. Since the interaction between the target and the ion-exchange medium is ionic the conductivity of the mobile phase is critical. Generally the conductivity of the feedstock is determined by the processes carried out upstream and significant modifications to the mobile-phase composition are not ideal owing to economic considerations. If the conductivity of the feedstock is too high for the ionic interaction between target protein and ion exchanger to occur then either the feedstock must be diluted with water or dilute buffer, or the pH moved further away from the pI in order to increase the charge on the protein. We have reported how the ionic concentrations and pH of the buffer affected the efficiency of chromatography of goat-serum proteins on Whatman QA52 [4] and demonstrate that the recovery and purity of the target IgG are affected by both of these factors for the reasons described. In a related study we investigated the effect of ionic composition of an egg-white protein feedstock on the binding efficiency of QA52 [5]. The egg-white feedstock was diluted with 0.02 M Tris/HCl buffer, (pH 7.5), which had a conductivity of ~1.8 mS. Using a 10 mg/ml egg-white protein feedstock the conductivity increased to ~2.75 mS and this was suitable for efficient adsorption to the QA52. However, when the egg-white protein concentrations were increased to ~20 mg/ml and ~30 mg/ml the conductivity increased to 3.44 mS and 5.59 mS, respectively. This had the effect of reducing the efficiency of binding of protein to the QA52 since the high conductivity of the feedstock caused a significant amount of the adsorbed protein to spontaneously desorb, leading to an inefficient adsorptive process, during a single pass through a column.

Adsorption/desorption kinetics of the medium
The adsorption/desorption kinetics of the ion exchanger are perhaps the most critical parameter influencing protein capacity, binding efficiency and chromatographic resolution. In order to ensure maximum throughput at low cost, it is necessary for as close to 100% as possible of the target to adsorb to the ion exchanger in a single contacting operation. This operation would typically take place in a column or in a batch stirred tank.

In either case the column effluent or the mobile phase following batch adsorption should be depleted of target such that the efficiency of the adsorption process is high and valuable target is not discarded as effluent. The adsorptive process will rely on mass transfer of the charged solute molecules to the external charged surfaces. If the pores of the ion exchanger are too small or inaccessible to the protein molecules then not only will the effective capacity of the exchanger be low but the kinetics of adsorption will be slow. If a medium is highly cross-linked to increase rigidity and support high flow rates, then the cross-links may restrict pore diffusion by generating intromolecular barriers which prevent a rapid adsorption process, between the solute and the ion-exchange groups.

Whatman ion-exchange celluloses have been demonstrated to exhibit very fast kinetics of adsorption and desorption for a range of proteins [6]. The kinetic uptakes of soybean trypsin inhibitor, bovine serum albumin and porcine thyroglobulin by DE52 at pH 7.5 are represented in Fig. 9.1. In terms of rates of adsorption/desorption in each case the time for 50% of the total protein to be adsorbed or desorbed ranged from 45 to 90 seconds. This data demonstrates the suitability of DE52 as a bioprocessing medium as it would be expected to exhibit good binding efficiency for the target within the contact time constraints of the adsorptive process.

Desorption kinetics are an important parameter in determining throughput parameters since these affect the resolution of the elution step and hence product purity. Desorption kinetics are of major importance for column chromatography where a linear salt or pH gradient is employed for elution of the bound proteins. In the case of anion-exchange chromatography, proteins would elute from the column in order of ascending acidity. It is at this stage that, by selection of an appropriate salt gradient and optimal fraction collection, the target will be obtained at the highest purity possible. Desorption of an individual component will occur when the concentration of counter-ion successfully competes with the solute molecule for the ion-exhange group of the medium. Once the solute molecule desorbs from the exchanger it must diffuse out of the pores rapidly and pass into the bulk flow of the mobile phase to elute from the column. If desorption kinetics are slow, i.e. the pore diffusion is impeded for the reasons discussed for adsorption, then conceptionally a weakly acidic protein which had desorbed from the inner surface of a pore would very slowly diffuse out of the pore. During this diffusion time the salt concentration of eluent could have increased such that a more strongly acidic protein adsorbed to the external surface of the ion exchanger would begin to desorb. The net effect would be co-elution of these two components resulting in poor chromatographic resolution and low product purity.

Gradient elution is the preferred technique provided that the desorption kinetics are fast. Where desorption kinetics are slow then a step elution may be employed but this will often lead to reduced product purity particularly where several proteins are adsorbed to the ion exchanger.

Process time

In order to make a process as efficient as possible, the incentive is to reduce process time, and where this involves column chromatography, increase the flow rate. The rigidity of the base support matrix determines the flow properties of an ion-exchange medium and also determines the maximum bed height which will support adequate flow at a realistic pressure across the bed. Being an adsorptive process, bed height is not too significant in ion-exchange chromatography and scale-up is generally achieved by increasing column diameter rather than increasing bed height. Whatman ion-exchange celluloses are often used at bed heights of 15–20 cm where adequate pressure/flow rate parameters are obtainable.

Although operating an ion-exchange process at high flow rate is attractive in principle, in practice the adsorption/desorption kinetics of the medium dictate the operational flow rate in order to obtain reasonable protein capacity and good chromatographic resolution. In the case of Whatman ion-exchnage celluloses a flow rate equivalent to two empty column volumes per hour, i.e. ~32 cm/h linear flow rate, is recommended. If a column were filled with buffer then at this flow rate the maximum residence time would be 30 minutes. Assuming a packed column of DE52 too contain 25% (v/v) DE52 then the maximum residence time of the mobile phase would be 22.5 mins. Clearly fast kinetics of adsorption would permit use of the majority of the total theoretical capcity in a single pass through the column whereas slow kinetics, particularly when the medium is partially loaded, woould give rise to early breakthrough and an inefficient adsorptive process.

In terms of desorption, slow kinetics give poor resolution, for the reasons discussed above, and this would be compounded with the use of high flow rates to even further reduce the chromatographic resolution of the medium. DE52 exhibits very fast kinetics of adsorption/desorption (Fig. 9.1) and for this reason is very effective at process-scale chromatography. In certain instances higher flow rates than

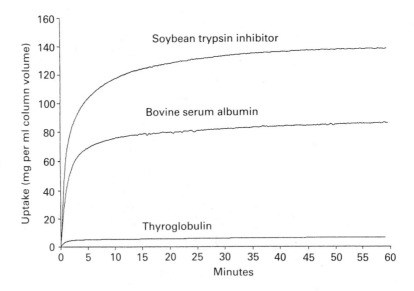

Fig. 9.1 — Kinetic uptake of soybean trypsin inhibitor, bovine serum albumin and porcine thyroglobulin by DE52 in 0.01M sodium phosphate buffer, (pH 7.5) in a stirred batch mode.

two empty column volumes per hour may be required, e.g. when dealing with a labile protein. In such circumstances the use of DE52 in a radial-flow chromatography column may be applicable. The use of this design of column enables DE52 to operate at flow rates up to fivefold faster than in a similar volume axial-flow column, with the fast kinetics giving rise to good chromatographic performance [7–9].

SCALE-UP OF THE PROCESS

Having assessed the requirements for the ion-exchange medium and developed the mobile phase to optimize the chromatographic step the process is ready for scale-up. There are two approaches to process-scale ion-exchange chromatography; these being the use of batch or column techniques. In simplistic terms a batch system is a simple equilibrium process whereby the ion exchanger is stirred with the feedstock in a large tank typically for about 60 minutes. A column process, on the other hand, is a continuous system whereby the ion exchanger is contained with a contactor fitted with porous bed supports to retain the medium, and through which the feedstock, etc., passes. Columns can use either axial or radial flow and these techniques have been reported elsewhere [7–9].

For process-scale ion-exchange chromatography the axial-flow column is of short bed height (15–20 cm) and large diameter, with scale-up effected by simply increasing column diameter whilst maintaining linear flow rate. For example, the Whatman Prep-25 column has an internal diameter of 45 cm and bed height of 16 cm to give a 25.4 l packed bed. This would be operated at a flow rate of ~0.8 l/min for DE52 (~32 cm/h) for the reasons discussed above.

The selection of a batch or column technique is influenceed by a number of factors including those summarized in Table 9.4. The nature and scale of the

Table 9.4 — Factors influencing the choice of batch vs. column techniques

(1)	Volume and concentration of feedstock
(2)	Scale, i.e. amount, of media required
(3)	Sanitary requirements
(4)	Media losses
(5)	Process time
(6)	Labour intensity
(7)	Centrifugation requirements
(8)	Shear constraints
(9)	Automation

upstream stages of the process will determine the volume and concentration of the feedstock for the ion-exchange step. Where a large volume of dilute feedstock is to be contacted with a small amount of ion exchanger then a batch process may be favoured, in order to reduce the prolonged column loading time if this technique were used. For example, treatment of 1000 l of feedstock using a 25 l column of DE52 would take ~21 h at a flow rate of 0.8 l/min compared with only 1–2 h using a batch technique. Batch may also be preferred where hundreds of kilograms of ion exchanger are being used owing to the limited availability and high cost of large columns. Batch techniques are, however, difficult to operate in an environment where sanitary conditions are required and a closed-column system may be favoured.

Being a contained system, media losses from a column are low compared with handling losses incurred during the batch operation. This may be a significant economical factor especially where expensive media are being used. Processing times are generally shorter in batch operations than in column operations. This will be dependent on feedstock volumes, elution volumes, flow rate, etc. Owing to the nature of the processes involved, batch techniques are more labour intensive than column, and require the use of a large filtration device to collect the ion exchanger from the batch stirred tank. This device is typically a continuous-flow basket centrifuge which is a major item of capital equipment and without which a batch process may be difficult to operate efficiently. Shear-sensitive media may be ineffective in a batch process owing to attrition caused by the stirring, pumping and centrifuging operations which are carried out in the batch operation. If the ion-exchange process is to be automated then a column approach is preferable and process control systems are available for automated column chromatography [10]. Having considered the points summarized in Table 9.4 and carried out an audit of the proposed process and the production plant capabilities the process would be ready for scale-up.

Comparison of batch vs. column techniques for process-scale anion-exchange chromatography

In order to compare batch and column techniques for process-scale anion-exchange chromatography the separation of hen egg-white proteins was carried out using Whatman DE52 anion-exchange cellulose. The investigation was carried out in three phases. Firstly, the chromatographic conditions were developed at laboratory scale and shown to scale up at least one-thousand-fold. Secondly, the maximum binding capacity of DE52 was determined using a suitable hen egg-white feedstock. Thirdly, the influence of batch vs. column techniques were examined for the process-scale separation of hen egg-white proteins.

Chromatographic conditions

A 10% (v/v) solution of cell debris remover (CDR)-treated hen egg-white containing ~10 mg/ml total proteins (7 ml) was applied to a column of DE52 (1.5 cm i.d. × 15.5 cm), pre-equilibrated with 0.025 M Tris/HCl buffer, (pH 7.5). Non-bound material was removed by washing with 0.025 M Tris/HCl buffer, (pH 7.5) (50 ml). Bound material was eluted using a linear gradient of 0–0.5 M NaCl in 0.025 M Tris/HCl buffer, (pH 7.5) (200 ml). The chromatography was carried out at a flow rate of 1 ml/min.

A parallel experiment was carried out using DE52 packed in a Whatman Prep 25 column (45 cm i.d. × 16 cm) at a flow rate of 1 l/min with a one-thousand-fold scale-up throughout.

The results of this study are represented in Fig. 9.2. Under analytical loadings, i.e. total protein loaded is <5% of total bed capacity, a typical separation of egg-white proteins was seen for DE52. The non-bound fraction represents the basic protein lysozyme, while the adsorbed proteins elute in the order conalbumin followed by ovalbumin. Having developed a suitable chromatographic system using the laboratory column (Fig. 9.2(a)) the separation was scaled-up one-thousand-fold

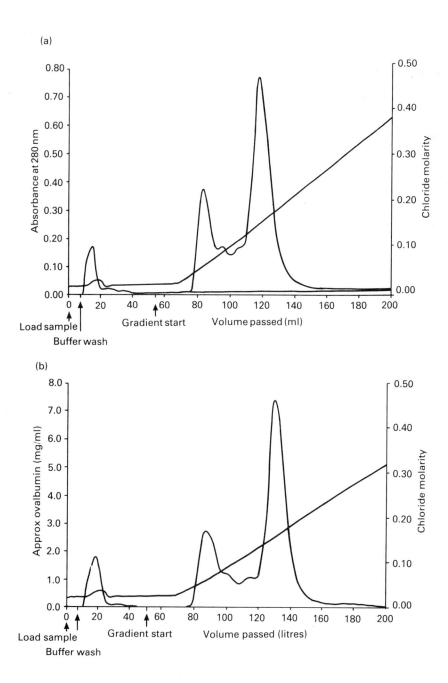

Fig. 9.2 — Column chromatography of hen egg-white proteins on DE52 using 0.025 M Tris/HCl buffer, (pH 7.5) at (a) laboratory scale (1.5 cm i.d.×15.5 cm) and (b) process scale (45 cm i.d.×16 cm).

by increasing the bed diameter thirty-fold whilst maintaining the column bed height and linear flow rate (34–38 cm/h). This separation (Fig. 9.2(b)) clearly demonstrates that the DE52 process has scaled-up satisfactorily.

Maximum capacity

A 10% (v/v) solution of CDR-treated egg-white was applied to a column of DE52 (1.5 cm i.d. ×15.5 cm) previously equilibrated with 0.025 M Tris/HCl buffer, (pH 7.5) at a flow rate of 1 ml/min until the absorbance at 280 nm of the eluate was similar to that of the feedstock. The absorbance profile of the DE52 column eluate is represented in Fig. 9.3. Following loading of 875 ml of feedstock (9.94 mg total

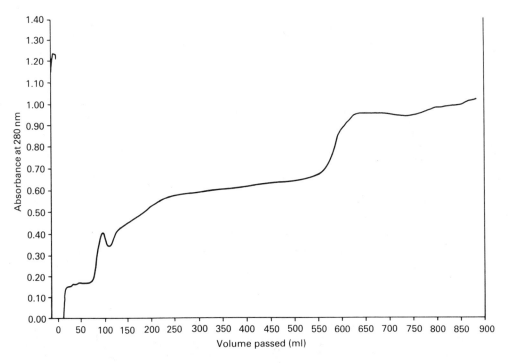

Fig. 9.3 — Absorbance profile of column eluate at 280 mm during saturation loading of DE52 with 10 mg/ml hen egg-white proteins using 0.025 M Tris/HCl buffer, (pH 7.5). Absorbance of the feeedstock is identified by the arrow.

protein/ml) the absorbance of the eluate resembled that of the feedstock (arrowed) indicating that a saturation loading had been achieved. During the adsorption stage a non-linear stepped absorbance profile was observed. This shape can be attributed to components of the egg-white eluting from the column in order of ascending acidity owing to protein:protein displacment, etc., resulting in the ovalbumin component only being retained. Similar observations have been reported for the treatment of goat serum with Whatman CDR (cell debris remover) [4]. Following column loading, and a buffer wash, a total of 4.34 g of protein was bound to the DE52 which reflects a capacity of 158 mg protein/ml packed-column volume. Elution of bound protein with NaCl resulted in a recovery of 98.6%.

In a process-scale separation, throughput is very critical and although it may be possible to operate a column of DE52 at a capacity of 158 mg/ml packed-column volume, this is not an entirely efficient adsorptive process since only 89.8% of the total ovalbumin present in the feedstock bound to the DE52, based on an average ovalbumin content of 63.8% (w/w) of total egg-white protein [11]. In separations where feedstock is freely available in excess then inefficient adsorption may be acceptable, but in many instances the value of the feedstock and/or target coupled with increased process times and added effluent treatment outweigh any potential benefits of such a step. It would therefore be desirable to carry out the process under conditions where as close to 100% of the target binds to the medium during a single contacting operation.

Process-scale chromatography
In order to compare batch and column techniques the separation of hen egg-white proteins using DE52 was carried out as follows:

600 eggs
(separate yolks)
20 l egg-white
(dilute)
10% (v/v) egg-white (200 l)
(clarify using Whatman cell debris remover (CDR))
post-CDR egg-white feedstock (200 l)
separation of proteins using 25 kg whatman DE52

The post-CDR egg-white feedstock contained ~8 mg total protein/ml, i.e. ~1.6 kg total protein contacted with the DE52. The DE52 chromatography was carried out using either batch or column adsorption and either batch or column desorption as summarized in Table 9.5. The column used for these studies was the Whatman Prep-25 (45 cm i.d. × 16 cm, 25.4 l). The experimental details have been reported elsewhere [12].

Table 9.5 — Chromatographic approach to process-scale separation of egg-white proteins on DE52

Adsorption	Desorption
Batch stirred tank	Batch centrifuge wall
Batch stirred tank	Batch stirred tank
Batch stirred tank	Column
Column	Column

In order to scale-up the chromatography whilst maintaining an efficient adsorptive process, 200 l of egg-white feedstock was applied to 25 kg DE52, i.e. a loading of

~64 mg protein/ml packed column volume, which is a submaximal loading as determined previously (Fig. 9.3). The loading capacity data for the process-scale studies are summarized in Table 9.6. The batch loading data indicates that ~70% of

Table 9.6 — Protein capacities during process-scale chromatography of egg-white proteins

					Total protein (g)		
Run	Adsorption mode	Desorption mode	Feedstock total protein (g)	Stage of chromatography	In mobile phase	Adsorbed to DE52	Binding efficiency (%)
1	Batch	Batch centrifuge wall	1479 — split into 2 lots	Loading wash elution	183 37 531	557 520 —	70.3
2	Batch	Batch stirred tank	after loading	Loading wash elution	187 36 444	552 516 72	69.8
3	Batch	Column	1574	Loading wash elution	427 61 1192	1147 1086 —	69.0
4	Column	Column	1703	Loading wash elution	263 119 1360	1440 1321 —	77.6

the total applied protein is adsorbed (binding efficiency) compared to a binding efficiency of ~78% for column loading. A column would be expected to be more efficient than batch since the former has several theoretical binding plates along its length whereas the latter, being a simple equilibrium process, can be regarded as having a theoretical plate count of one. The salt elution in a column operation gave ~100% recovery of bound protein. In a batch stirred tank, elution of bound protein was incomplete, presumably owing to the equilibrium nature of this simple process. Elution of bound protein from the wall of the rotating basket centrifuge was very efficient presumably since the pad of DE52 on the centrifuge wall represents a radial-flow column, which was continuously eluting and therefore would have a theoretical plate count of greater than one. The data indicates that ~53 mg proteins bound per ml(g) DE52 following a loading of ~64 mg total potein/ml. Based on an ovalbumin content of 63.8% (w/w) of total egg-white protein [11], then assuming 100% adsorption of the ovalbumin (~41 mg/ml), only small amounts of other acidic proteins would be expected to have bound to the DE52. This was found to be the case when the adsorbed material was assayed for homogeneity [12].

The data summarized in Table 9.6 suggests that column steps are the most efficient both in terms of adsorption and desorption, but they are both lengthy processes and will be restricted by the column flow rate for the reasons described earlier. The labour times for the individual process stages of the batch and column chromatography are summarized in Table 9.7. These data indicate that batch

Table 9.7 — Labour times for individual process stages of batch and column chromatography of hen egg-white proteins on DE52

Stage of process	Labour time (man hours)			
	Batch adsorption Batch desorption (stirred tank)	Batch adsorption Batch desorption (centrifuge wall)	Batch adsorption Column desorption	Column adsorption Column desorption
Adsorption (Batch)	2	2	2	–
Column packing	–	–	1.5	2
Adsorption (Column)	–	–	–	4
Buffer wash	1.5	1.5	2	2
Salt desorption	3.5	2.5	4	4
Overall process time	7.0	6.0	9.5	12.0

adsorption/desorption stages are faster than the comparable column operations and where larger volumes of feedstock are used then the batch separation would become even more time efficient.

MEDIA RE-USE

In many instances, the ion-exchange medium is used many times and whilst ionically bound proteins, etc., can be eluted using a high-ionic-strength eluent, media fouling often occurs where unidentified materials remain retained by the media during desorption and this affects capacity and performance during subsequent use. In such a situation a stringent regeneration protocol must be employed to restore chromatographic performance. We have developed clean-in-place (CIP) protocols which have been demonstrated to be effective in regenerating columns of QA52 [4] and DE52 [13]. Following four consecutive process-scale loadings of QA52 with goat serum, the capacity for IgG reduced anbd chromatographic resolution was impaired [4]. Following five consecutive process-scale loadings of DE52 with hen egg-white, the capacity of the medium was relatively unaffected but the chromatographic performance of the bed was impaired [14]. In each case treatment of the bed of QA52 or DE52 with 0.5 M NaOH for 12 hours restored the column performance to that observed after the first run [4,14].

Procedures for CIP of Whatman anion-exchange celluloses are described elsewhere [13,14] although media regeneration is process-dependent and it is impossible to give a generic CIP protocol which is effective in all applications. However, in our experience the minimum requirement is 0.5 M NaOH, but in certain instances an ionic strength of up 2.0 may be required at elevated temperatures of at least 60°C. The increased ionic strength may be achieved by increasing the concentration of NaOH or by addition of NaCl.

In terms of stability of the ion-exchange medium to prolonged storage in NaOH, we have demonstrated that DE52 and QA52 are stable in up to 2 M NaOH at temperatures of up to 42°C for 3 months [13]. Over this period of exposure to NaOH, the regains of the media (measure of the degree of swelling), small-ion capacity and capacity for bovine serum albumin were not significantly affected. These data are summarized in Table 9.8.

Table 9.8 — Stability of DE52 and QA52 in 2 M sodium hydroxide at 42°C (batch storage)

Grade	Small-ion capacity (meq/dry g)		BSA uptake (g/dry g)		Regain (g/dry g)			
					NaOH washed		HCl washed	
	Sample	Control	Sample	Control	Sample	Control	Sample	Control
QA52								
1 week	1.14	1.23	1.02	0.97	3.05	2.95	3.27	3.15
1 month	1.23	1.23	1.01	0.97	3.09	2.95	3.33	3.15
3 months	1.24	1.18	0.96	0.94	3.31	3.06	3.47	3.25
DE52								
1 week	1.01	0.99	0.737	0.707	2.56	2.49	3.04	2.91
1 month	0.98	0.96	0.737	0.742	2.64	2.61	3.17	3.11
3 months	0.98	0.99	0.795	0.721	2.67	2.59	3.30	3.08

MEDIA SANITIZATION

In certain manufacturing processes a pre-requisite for the ion-exchange step would be that it is carried out under sanitary conditions. In general this process would be carried out in a column and, following use, the column and its contents would need to be sterilized and depyrogenated prior to a subsequent step. The procedure developed for CIP of Whatman ion-exchange celluloses [13,14] has been shown to be efficacious at simultaneously sterilizing and depyrogenating columns of CDR, DE52 and QA52 [15,16]. In these studies, laboratory columns of CDR, DE52 or QA52 were challenged with a suspension of *Escherichia coli* such that the column effluents contained ~1×10^9 viable organisms/ml and were highly pyrogenic. Following storage in 0.5 M NaOH for 12 hours and re-equilibration with sterile pyrogen-free buffers, the column effluents were sterile, pyrogen-free and contained insignificant levels of endotoxin.

CONCLUSIONS

In any ion-exchange process, the initial chromatographic system will be developed at laboratory scale. It is critical that correct media selection is made at this stage and factors such as capacity for the target, adsorption/desorption kinetics, flow performance, protocols for regeneration and sanitization, etc., should all be considered. The process would then be scaled-up typically using a batch or column approach. In terms of productivity a batch technique may be less efficient than using a column but the former may be less time-consuming and costly which may be important considerations. Since the commercial success of an ion-exchange process will be influenced by many of the factors discussed in this review a critical assessment of the proposed process should be made prior to scale-up to ensure commercial viability.

REFERENCES

[1] L. Stryer, *Biochemistry*, 2nd edn, W. H. Freeman & Co, San Francisco, p. 11 (1981).
[2] D. Freifelder, *Physical Biochemistry*, 2nd edn, W. H. Freeman & Co, San Francisco, p. 249 (1982).
[3] E. F. Rossomando, *Methods in Enzymology*, **182** (ed. by M. P. Deutscher) Academic Press, San Diego, p. 309 (1990).
[4] P. R. Levison, M. L. Koscielny and E. T. Butts, *Bioseparation*, **1**, 59–67 (1990).
[5] P. R. Levison, D. W. Toome, S. E. Badger, B. N. Brook and D. Carcary, *Chromatographia*, **28**, 170–178 (1989).
[6] P. R. Levison, D. W. Toome, S. E. Badger, D. Carcary and E. T. Butts, *Pittsburgh Conference Abstract Book*, New York, p. 1079 (1990).
[7] L. Lane, M. L. Koscielny, E. T. Butts and P. R. Levison, *Pittsburgh Conference Abstract Book*, New York, p. 1078 (1990).
[8] L. Lane, M. L. Koscielny, P. R. Levision, D. W. Toome and E. T. Butts, *Bioseparation*, **1**, 141–147 (1990).
[9] P. R. Levison, *Pharmaceutical Manufacturing International* (1991) in press.
[10] G. K. Sofer and L. E. Nystrom, *Process Chromatography, a Practical Guide*, Academic Press, San Diego, p. 67 (1989).
[11] W. Bolton, *Biochemist's Handbook*, (ed. by C. Long, E. J. King and W. M. Sperry) E. and F. N. Spon Ltd, London, p. 764 (1971).
[12] P. R. Levison, S. E. Badger, D. W. Toome, D. Carcary and E. T. Butts, *Advances in Separation Processes*, Inst. Chem. Eng. Symp. Series, **118**, pp. 6.1–6.11 (1990).
[13] Whatman Biosystems Ltd, *Clean-in-Place Procedures and Re-use of Whatman Cellulosic Media*, Leaflet IEC 4 (1988).
[14] P. R. Levison, S. E. Badger, D. W. Toome, D. Carcary and E. T. Butts, *Separations for Biotechnology*, **2**, (ed. by D. L. Pyle) Elsevier Applied Science, London, pp. 381–389 (1990).
[15] P. R. Levison and F. M. Clark, *Pittsburgh Conference Abstract Book*, Atlanta p. 754 (1989).
[16] P. R. Levison, S. E. Badger, D. W. Toome, D. Carcary and E. T. Butts, *Downstream Processing in Biotechnology* **II** (ed. by R. de Bruyne and A. Huyghebaert) The Royal Flemish Society of Engineers (k viv) Antwerp, pp. 2.11–2.16 (1989).

10

Continuous adsorption and chromatography in the purification of fermentation products

Gordon J. Rossiter
Advanced Separation Technologies Inc., Lakeland, USA

ISEP† CONTINUOUS MOVING BED CONTACTOR

Adsorption separation processes are based on the sequencing of solid sorbents through a cycle of specific steps in each of which the solid sorbent is contacted with a fluid phase. Through modification of the fluid phase properties and composition the sorbent will adsorb an adsorbate in one step (adsorption) and release or desorb the adsorbate species in a subsequent step, typically called regeneration or stripping. Then the cycle recommences with the adsorption step although often in liquid processes there are also sorbent wash, backwash and rinse steps in a complete cycle. Traditionally, adsorptive separation processes are conducted in fixed beds.

Fixed-bed adsorption systems have traditionally been applied to processes that require the removal of dilute concentrations of chemical species. The low concentrations permit beds of practical size to remain in service for long periods of time — 4 to 24 hours. The major disadvantage of fixed bed sorption systems are the inefficient use of sorbent, the excess consumption of regenerant chemicals and the limited applicability to only low-feed concentrations. The continuous systems overcome these obstacles through counter-current processing techniques that improve the contacting efficiency. The ISEP machine in addition provides a new dimension in design flexibility that addresses some of the more serious drawbacks to using packed beds in industrial separations.

† ISEP (Ion SEParator) is a patented device marketed by Advanced Separation Technologies (AST).

Ch. 10]   Continuous adsorption and chromatography

The ISEP is a patented device that implements cyclic sorbent-based separation processes in a continuous manner while removing the constraints and inefficiencies of the fixed-bed designs. The ISEP has been designed with process flexibility in mind. Specifically, ISEP continuous operation brings the following advantages:

(1) continuous steady-state, non-interrupted process flows enter and exit the ISEP,
(2) decreased sorbent inventory requirements, as all sorbent is active in a zone,
(3) higher desorption product fluid (eluate) concentrations from countercurrent contacting,
(4) higher eluent efficiency, decreased eluent usage from countercurrent contacting,
(5) high tolerance to suspended solids in feed liquids, upflow contacting possible,
(6) minimized wash-solution requirements resulting from countercurrent operation,
(7) zone lengths are individually optimized, connecting beds in series,
(8) zone contact areas are variable, connecting beds in parallel.

Fig. 10.1 (ISEP GA) is a general arrangement drawing of an industrial ISEP machine. Important component parts are identified in the drawing legend. The discussion in this paper is limited to the AST 'standard ISEP' configuration. This is a 20 fixed port/30 chamber ISEP unit. In point of fact, the numbers of fixed and rotating ports need only observe the relationship,

$$\text{Number of rotating ports} \equiv \text{chambers} > \text{number of fixed ports}$$

From a functional standpoint the important ISEP components are:

(a) the two valves, each with 20 fixed ports and,
(b) the 30 chambers, each with one connection to each valve.

The valves are mirror images of each other and serve to accept the fluid inputs to, and fluid outputs from, the process through 20 fixed ports on each valve — upper and lower. The valves distribute the various fluids to the 30 sorbent chambers. Because the ISEP valves are the heart of the system the following section will discuss the valve geometry.

Valve geometry
The patented valves have 20 fixed ports and 30 rotating outlets. The valve geometry is illustrated in Fig. 10.2. There are two basic rules that must be obeyed in designing this valve and they are, (1) a fixed port must *always* have access to at least one rotating outlet which guarantees fluid access to sorbent at all times, and (2) a rotating outlet (connected to a single sorbent chamber) must *only ever* be connected to a single pair of upper and lower ports. A properly operating valve will have a perfect alignment of ports and rotating outlets.
    Referring to Fig. 10.3, we define the following dimensions,

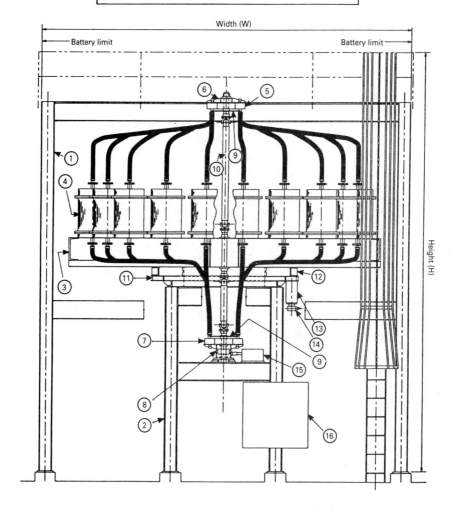

Fig. 10.1 — Standard arrangement diagram of an industrial ISEP machine.

Ch. 10] Continuous adsorption and chromatography 165

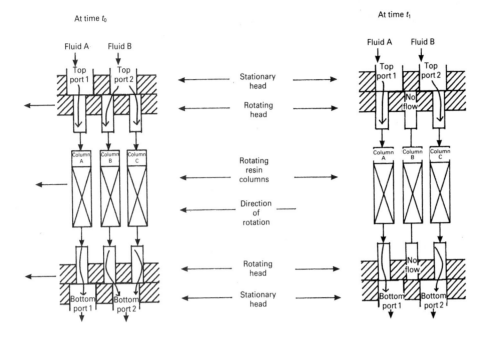

Fig. 10.2 — ISEP valve geometry.

n_p = number of fixed ports,
n_s = number of rotating ports, 'slots' or outlets ≡ number of chambers,
W^o = width of a port or 'window' on the fixed plate of the valves in degrees,
w^o = width of the web seal between port openings on the fixed plates in degrees,
s^o = width of a 'slot' or opening on the rotating plate of the valves in degrees,

then for a 20/30 valve (20 ports/30 chambers),

$W^o + w^o = 360/n_p = 18°$, and there are $360/18 = 20$ zones like this,
$d_s^o = 360/n_s = 12°$ centre-to-centre distance between adjacent rotating outlets,
$w > s$, web must be wider than the width of the rotating outlet in order for the chamber to seal off as it switches between adjacent ports,
$w^o = 2s^o$ typically and,
$W^o > (360/n_s - s^o)$, 'window' must be wide enough so that it always 'sees' a 'slot'.

In a typical valve $W = 12.5°$, $w = 5.5°$ and $s = 2.5$–$3.0°$.

From consideration of these relationships one can maximize or minimize the on-stream time of a 'slot' or resin chamber. For example, in adsorption processes where selectivity is high and few contact stages are required, then maximum on-stream time

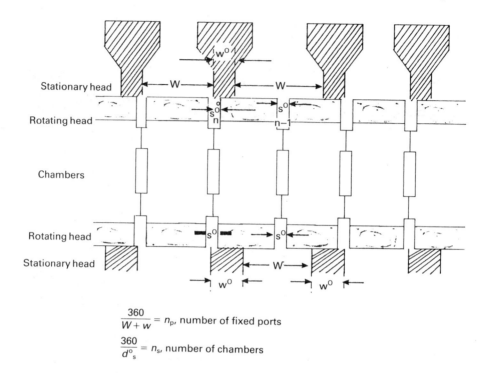

$$\frac{360}{W+w} = n_p, \text{ number of fixed ports}$$

$$\frac{360}{d^\circ_s} = n_s, \text{ number of chambers}$$

Fig. 10.3 — ISEP valve geometry.

is desirable and W and/or s is set at its widest. On the other hand when chromatographic fractionations are implemented between species of similar sorbent affinities, we wish to minimize the mixing effects that occur when two chambers are sharing the same port. In such cases the width dimensions for W and s are now reduced to a minimum consistent with acceptable pressure drop. Deliberate misalignment between the two valves can accomplish the same result as decreasing s. Acceleration of the drive on a timed basis when chambers are making the transition of entering the exiting 'windows' also maximizes the versatility of the ISEP valving system.

The physical size of the valve is a function of the size of dimension S, the slot width in centimetres. This is dictated by the expected maximum flowrate ($m^3 hr^{-1}$) through a single port. The slot width is then chosen so as best to contain fluid pressures within the ISEP system to values below the rated operating pressure. Pressure is necessary to drive the fluids through the valve passages, the ISEP interconnecting piping and through the sorbent beds. As a design guideline, the maximum pressure drop through a single valve should be held to 0.5 bar. ISEP-valves are designed to operate at 2 to 2.5 bar. The required s value, once determined, will set the diameter, D, of the ports in the valve.

$$D = (360/s) \times S/\pi,$$
e.g. for $s = 3$, $S = 1.2$ cm, then $D = 45.8$ cm.

The finished valve size is larger owing to the radial seal distance of 1.5–2.0 cm and other considerations such as the room needed to service external fittings attached to the valve.

Chambers

Each of the 30 ISEP chambers is identical in all respects. In the operation of the ISEP it becomes important to maintain this equality. A chamber drawing is shown in Fig. 10.4. There are many different approaches to chamber design but each one normally

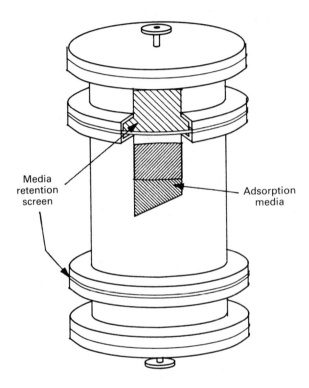

Fig. 10.4 — Cylindrical resin cell.

addresses the following aspects:

— (1) aspect ratio ≡ [column length/column diameter],
— (2) freeboard volume for bed expansion during upflow/backwash,
— (3) screen design — lower screen not only prevents sorbent from escaping but must also be strong enough to support the bed weight,
— upper screen is designed to stop losses of sorbent during the fluidization steps. It also serves as part of the downflow distribution system,
— (4) good fluid distribution,

— (5) access for sorbent and chamber inspection,
— (6) ability to rapidly charge and discharge sorbent when necessary.

The aspect ratio or $L:D$ value concerns the bed depth (L)–bed diameter (D) relationship and it is recommended that,

$$L:D \geqslant 1.0$$

This is an experience factor that comes from engineering practice with packed fixed beds. In any naturally packed bed there will be a tendency to channel, to operate without a truly flat upper surface, etc. Choice of a design $L:D$ ratio$\geqslant 1.0$ minimizes such effects. Note that the aspect ratio of the laboratory column tests is $>>>1$ and is typically of the order 20–40. *Aspect ratio is the major scale-up parameter* in ISEP design.

The amount of freeboard needed above the bed is case-dependent. ISEP designs usually incorporate at least one 25% bed expansion step (1 port minimum) in all designs so that beds are 'turned over' once per revolution thereby eliminating any tendency for channelling to occur. Designs for 50%, and more, expansion may be necessary in cases where higher upflow velocities are desired, severe solids accumulation results if backwashing is not performed thoroughly, etc.

Screens consist of filter-cloth type mesh (mesh size finer than sorbent size), or wedgebar wire screens with the properly sized openings. The former require additional support which is supplied via perforated plates that act as a mount for the screen-cloth material. The upper screen often incorporates a 'target plate' on its upper side to disperse the incoming fluid jet and dissipate its downward-directed momentum. If not checked, it is possible that a fluid jet would penetrate the upper screen and gouge a hole in the top of the sorbent bed.

Good fluid distribution comes from proper attention to the screen and fluid inlet designs. If the bed-top condition can be maintained fairly flat, good distribution of fluid flow through the bed will be assured. The once-per-revolution back-flush also ensures homogeneous conditions within the bed which in turn provides for a good even fluid velocity profile across the bed diameter.

Cell design often incorporates a special nozzle for resin bed service. This service nozzle can be used to sample resin, inspect the cell interior and is used to hydraulically charge and remove sorbent from the chamber.

Comparison with fixed-bed operation
Consider a fixed bed of solid sorbent as shown in Fig. 10.5. Such a bed of sorbent is typically contained inside a cylindrical vessel supported on a bed of coarse media or resting on a screen. There is a freeboard volume present above the bed to accommodate bed expansion during backwashing. The diagram in Fig. 10.5 shows the mass transfer profile in this bed at two points in time t_1 and t_2 where $t_2 > t_1$. The cross-hatched area above the profile at t_1 represents the exhausted resin inventory at time t_1. Likewise at time t_2 the area below the profile corresponds to as yet unexchanged or regenerated sorbent. Note that the horizontal axis is used to

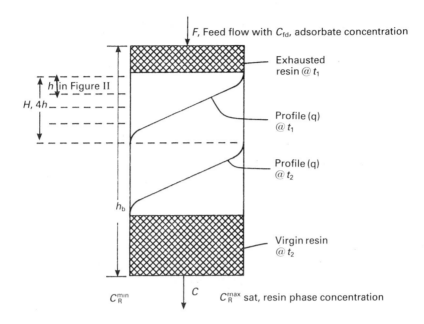

(1) Stable profile established at time t, with a mass transfer zone (MTZ) length of $4h$, $= H$

(2) Profile or front travels through the bed. $v_{front} - 4\,h / (t_2 - t_1)$
$- F * C_{fd}/(C_R^{max} - C_R^{min})$

(3) The MTZ profile is characteristic of the system and depends on the sorbent, flowrates, the sorbable species, exchange kinetics etc.

Fig. 10.5 — Mass transfer profile: fixed bed.

represent $[q]$, the adsorbate concentration in the solid sorbent phase. The vertical axis is simply the bed height. The rate of travel for this mass transfer front is:

$v_{front} = 4h/[t_2 - t_1]$, where $4h$ = MTZL, $t_2 - t_1$ = time for MTZL move $4h$,

$= R_{flow}/A$, where R = resin flow or exhaustion rate, A = bed area,

$= F \times [C_{feed}/C_{loaded\,resin}]/A$, F = feed flowrate, C = g/l concentration

It becomes immediately clear that in a fixed-bed batch-operation there is, by the very definition, more sorbent inventory than is contained in the active zone of mass transfer. The amount of this extra inventory is determined by the system requirements for the non-adsorption steps in the process. In contrast a continuous system is designed to make the sorbent flow in the opposite direction to the travel of the mass transfer profile and at the correct velocity such that the profile remains stationary in space. All continuous systems attempt to accomplish this objective.

Sorption separation systems include a step of sorbent regeneration. In fixed-bed systems this operation and other off-line steps such as washing and rinsing require time. The time between removing two consecutive beds from the adsorption step is called the system cycle time, T_{cycle}. During this time the fixed-bed designer has two choices. He can store the feed liquid containing the adsorbate or he can, as is normally the case, design enough capacity in the adsorption sorbent inventory such that the mass transfer profile does not exit the beds in service during this period of regeneration for the offline bed. This inventory or 'in service' resin is,

$$\text{Resin volume required in service} = T_{\text{cycle}} \cdot v_{\text{front}} \cdot A + \text{MTZL}$$

In the case of the continuous systems the time for regeneration and sorbent inventory in service are independent and can be optimized separately. ISEP design procedures exist to determine each zone's residence time requirements.

Looking at Fig. 10.6 we see how in the ISEP continuous system we divide up the total length of bed necessary for mass transfer into several segments.. Note that in Fig. 10.5 we divided the mass transfer profile into four 'slices', each of height h where

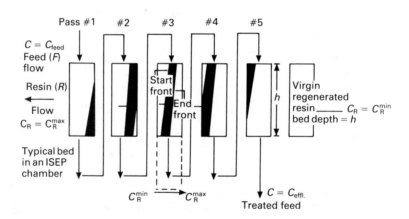

(1) Mass balance $F \times [C_{\text{feed}} - C_{\text{effl.}}] = R \times [C_R^{\max} - C_R^{\min}]$

(2) Number of passes $= n = H/L\ h + 1$.

(3) Shaded area represents the movement of the adsorption front, left to right, in a chamber while it is in a particular pass.

Fig. 10.6 — Mass transfer profile : ISEP.

$4h = H$, the mass transfer zone length (MTZL). Back to Fig. 10.6 each chamber in the ISEP machinee contains a short sorbent bed of height h. Typically $h = 0.3$–1.2 m depending on the process. An aspect ratio (bed depth (h) : diameter (d)) of 1.0 or greater is usually designed. Again, as in the large chamber drawing notation, we've used the horizontal axis on each chamber to represent $[q]$, adsorbate concentration in the sorbent phase. The 'mini profiles' shown in each chamber at time t_1 and t_2 are short sections of the sloped profile in the large fixed bed of Fig. 10.5.

The chambers shown in Fig. 10.6 are moving at a fixed rate from right to left† through the adsorption zone and spend a fixed amount of time in each of five sub-zones. The adsorption zone consists of five ports or sub-zones connected in series fashion so as to provide a path for the fluid through a bed $5h$ deep. As a chamber moves beneath a port it receives fluid flow and the sorbent commences or continuous adsorbing adsorbate. The rate of resin chamber movement is determined by the designer and is a function of the average resin flow rate, R. In Fig. 10.5, at the moment a chamber enters the sub-zone beneath a port its sorbent concentration profile is represented by the left-hand extremity of the dark band shown inside each small bed. At the moment a chamber exits the port in question the profile has migrated to the right-hand extremity on the dark band. The initial profile for a chamber arriving from regeneration corresponds to a vertical line on the left-hand edge which represents a sorbent in the fully regenerated form where the whole bed is at $[q]_{min}$. The ideal ending profile should represent a sorbent saturated with adsorbate and will be a vertical line at the right-hand edge of the chamber rectangle where all sorbent is at concentration $[q]_{max}$.

A couple of observations are worth making at this point with respect to the efficient use of sorbent in this arrangement of the mass transfer zone. First, the sorbent chamber in contact with new feed will have all its resin saturated to an equilibrium level with the feed if the process design calls for such a condition. Since we normally wish to maximize the sorbent capacity, the leading (leftmost) chamber must exit port #1 saturated. At the moment the fluid composition entering port #2 will contain $[c]_{feed}$; that is the mass transfer zone start point is now a distance h into the $5h$ total bed. For this reason we require $5h$, in this five port case, to always contain the mass transfer zone length within the adsorption zone resin bed. This leads us to the obvious conclusion that it is advantageous to divide the MTZL into as many small segments as is reasonable.

if H = MTZL determined from fixed-bed testing (@ specified velocity), then
if h = height of ISEP shallow beds, then
$n = H/h + 1$, where n = number of passes/subzones in the ISEP adsorption zone.

As n increases $n/n + 1$ approaches the ideal 1.0. The limitation to employing large numbers for n are twofold:

(1) There are cost limitations to a total number of ports available in an ISEP machine and 20 ports is the standard. Ports are required for the steps of washing, regeneration, rinsing as well as adsorption. Passes through the bed in a single

† ISEP resin travel direction from right to left is an AST standard convention.

zone range from 1 to 8 in AST experience. Twenty ports has normally satisfied most of the configurations in which the ISEP is applied.

(2) A certain minimum bed depth of 0.2 m is advisable in order to maintain a good aspect ratio and fluid distribution. If the MTZL < minimum bed depth then the segmentation does not apply. *Later the MTZL variation with fluid velocity is treated and often short MTZLs can be deliberateley lengthened to decrease bed area needs and reduce machine capital costs.*

The ISEP unit has an advantage that deserves mention in the chemical processing arena where tolerance to solids is important. Operation of a 15 m^{3}h^{-1} pond water flow containing up to 1000 ppm suspended solids worked trouble free during the one year campaign to remove fluorine from phosphoric acid pond waters. Fig. 10.7 illustrates the kind of flexible configuration that might be used to combine the efficiency of the packed-bed series adsorption with a following parallel upflow wash zone long enough to backwash solids from the resin media.

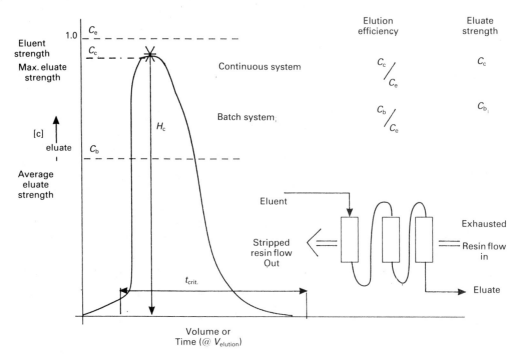

Fig. 10.7— Continuous vs. batch elution.

Elution efficiency in countercurrent systems comes from operating the elution at a single point on the elution profile curve shown in Fig. 10.8. If one imagines a column of exhausted resin that is put into regeneration and the eluent peak begins to emerge from the column effluent; now imagine moving the resin in the opposite direction at such a rate as to maintain the peak effluent level. This is how the ISEP elution zone is designed to work.

Continuous adsorption and chromatography

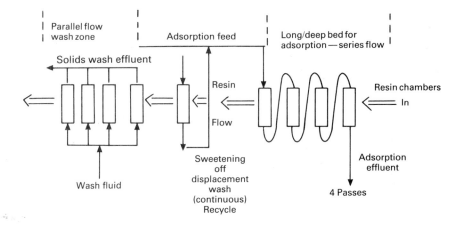

(1) Shallow beds purge of solids — less time required
(2) Controlled flow in expanded beds.
(3) Allocate required sorbent residence time.

Fig. 10.8 — Solids tolerance in shallow beds.

A typical schematic diagram of a fixed-bed operation is shown in Fig. 10.9. For any mildly complex process (number of steps > 4) the amount of piping, tankage and valving necessary to service such a system is complicated, expensive and is built up of many parts that can fail and shut down the operation. Each valve is actuated (controlled) using position detectors all of which can cause the system to malfunction. The solution heel in the lines between valve and bed can cause design problems when chemicals are incompatible which increase the cost. The bed cycle time is often designed around a non-technical factor which causes the system to be larger than an optimum size. For example, one regeneration/shift is stipulated. In other cases the cycle time is once per day since the regeneration must be done during the day when qualified technicians are available. The ISEP suffers from none of these restrictions since it operates like any continuous operation and requires no more normal attention than a process pump or filter.

ECONOMIC COMPARISON

Large reductions in sorbent inventory result from proper design of continuous ion exchange (IX) systems like the ISEP. Basically, as explained above, the ISEP design uses the resin inventory usefully at all times. The idle inventory in fixed-bed systems has two origins; the nature of fixed-bed adsorption in which a zone travels through the bed for a fixed period of time; and, the consideration given to the operator attention required at specific intervals, e.g. once per shift per day.

An example of the economic impact of dramatic savings in sorbent inventory can be appreciated by examination of the following table excerpted from two economic

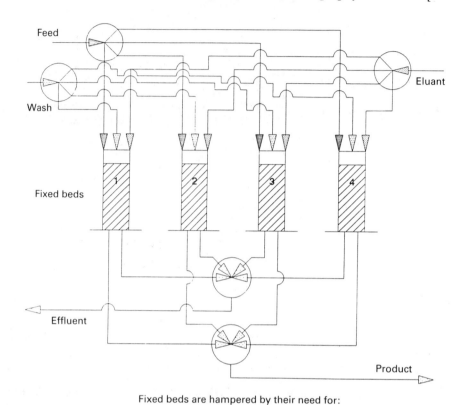

Fixed beds are hampered by their need for:

Multiple large sorbent beds

Complicated timing and switching of many valves

Interruption of feed and eluant flows

Constantly varying outlet concentrations

Inefficient use of eluant and wash water

Fig. 10.9 — Diagram of fixed-bed operation. Standard technology for solid a sorbent/liquid contact is a fixed-bed.

comparisons for two different processes. The lysine case is elaborated on in further detail later. The highlights as far as sorbent savings go are summarized in Table 10.1.

Table 10.1

Process	ISEP	Fixed bed	$/m^3	Total savings
Lysine purification	32 m^3	549 m^3	8 800	4 555 000
Cobalt recovery	85 m^3	652 m^3	28 000	15 876 000 (USBM)

The approach to fixed-bed design is to choose an acceptable cycle time. The cycle time is defined as the length of time between removing consecutive beds from service. This cycle time is usually a function of the time required to treat an exhausted column with regenerant wash and other fluid operations. The cycle time can be cut short if more than one column is assigned to the 'off-line' or 'out-of-service' tasks. This only increases the sorbent inventory and complicates the system piping and controls. The resin inventory 'in-service' must be sufficient to adsorb all the production that corresponds to the cycle time and in addition contains the mass transfer zone within the bed. An example calculation illustrates this more clearly (see Table 10.2).

Table 10.2

Process: lysine broth purification, lysine feed 100 g/l
MTZL: 3 m
Resin capacity: 120 g/l lysine.HCl
Feed flow: 25 $m^3 hr^{-1}$
Cycle time: 4 hours [includes washes (1 bed), backwash (1), stripping (2)]
Adsorption flow: 1.5 mhr^{-1}
Beds in system: 3 adsorption, 3 stripping, 2 washing (1 after strip, 1 after adsorption)

Resin exhaustion rate	= $(110/120) \times 25 = 22.9$ $m^3 hr^{-1}$
Inventory in a single bed	= $4 \times 22.9 = 91.6$ m^3
Inventory in 6 bed system	= 549 m^3
Diameter of fixed bed	= 4.60 m
Height of bed	= 5.50 m, aspect ratio $L:D = 1.2:1$
Adsorption beds	= 2
Total beds/columns	= 6

Note that the choice of $L:D$ has meant that the mass transfer zone (MTZL) is contained within a single-bed depth. The commercial ISEP installation for this same service employs only 32 m^3 of sorbent. Table 10.3 summarizes the economic analysis.

Table 10.3 — Fixed bed vs. ISEP lysine purification systems

	Fixed bed	ISEP
Capital cost (US$)	1 600 000	1 500 000
Resin initial charge (US$)	4 555 000	128 000
Total	**6 155 000**	**1 628 000**
Product concentration (g/l)	80–90	180–200
Ammonia eluent %efficiency	30–35	65–70
Product recovery %	85–90	95 min

The basis for fixed-bed costing was the cost of a single stainless-steel vessel with welded internals, external legs and the necesary inlets and outlets. This cost was then factored (five times) to estimate the installed cost. The valving, programming and control, piping and installation labour is included in this factor. The ISEP cost likewise contains field costs, associated instrumentation, and tanks and pumps.

The improved product concentration and ammonia eluent usage come directly from the countercurrent operation possible in the ISEP elution and washes. Higher recovery stems from the unique ability of the ISEP to modify solution pH within the adsorption zone at a given point. One can see from this analysis that even if the fixed bed was optimized further, it could never compete with the ISEP continuous contactor in the lysine purification application.

The differences shown here between fixed-bed and continuous IX become less dramatic as the feed solution becomes more dilute. Still, as solutions become more dilute — [$C_{feed} < 0.5N$] — we have found that in many cases the adsorption zone length requirement is described by the following relationship:

$$\text{MTZL} \; \alpha \; v^{0.5}, \text{ where } v = \text{fluid superficial velocity} = \text{flow rate/bed area}$$

Consequently, a doubling of the specific flow rate (fluid velocity) will require only 40% more bed. The ISEP employs shallow beds of depth < 1 m which do not have large pressure drops. Flow rates upwards to 25–50 m^{3}hr^{-1} are being designed into commercial systems.

The other attractive aspect of ISEP technology is the enormous flexibility of sorbent inventory allocation. In systems where adsorbate concentration is low, up to 75% of the machine is dedicated to adsorption. A combination of high specific flows and flexible allocation of total resin inventory is making ISEP a viable choice even in areas of conventional fixed-bed applications.

ADSORPTION PROCESS DESIGN PROCEDURES
General
Laboratory testing, lasting in some cases only as long as a couple of weeks, can provide sufficient information to demonstrate the chemical feasibility of the separation, establish the size of the various mass transfer zones and indicate the efficiency of each contact. This information can accurately define the capital and chemical consumption quantities suitable for conceptual economic project analysis.

The testwork required to reliably size a continuous adsorption ISEP unit is remarkably simple and involves standard column tests and equilibrium contacts. Resin or sorbent selection is a first step that relies on experience and screening selected candidates using the column and contacting tests described below.

Column adsorption and elution testing is typically performed at various velocities and with different bed depths to provide the basic mass-transfer kinetic data. Column-effluent samples are taken at appropriate volume intervals and analysed for the species of interest. The effluent analytical data is plotted as $C_{effluent}$ vs. V (cumulative), where V is in units of millilitres or column bed volumes, while C is typically in units of grams/litre.

Batch contacts, at controlled temperature, with time removal of samples, provides further kinetic data on approach to equilibrium rates and equilibrium data at various solution compositions. The isotherm data are plotted as C_{resin} vs. C_{liquid}, where C is usually in units of grams/litre. The timed approach to equilibrium data are plotted as C_{liquid} vs. T (minutes). This data is useful in modifying the interpretation of isotherm data.

Equilibrium data treatment and generation

Isotherms are relationships that describe the phase distribution of a component at some fixed temperature. In our IX work we need to know the phase distribution for an absorbable species across the range of conditions it sees during its passage through a zone of mass transfer in the ISEP. Thus relationships are referred to as K_d values and are plotted on x–y graphs with $[C]_{liquid}$ on the x-axis and $[C]_{resin}$ on the y-axis.

The experimental techniques to generate this information are simple and not costly. In the case of a feed solution for instance the experimenter first prepares a batch of resin (1 litre) in the fully regenerated form that the proposed process calls for. Then 100 millilitre fluid batches are placed in separate beakers. Then different amounts of resin are added to each beaker. These amounts might range from 100 millilitres down to 5 millilitres. After equilibration the various liquids are sampled, analysed and the calculations made to plot the isotherms.

Stripping isotherms are generated in identical fashion where the resin batch prepared is the exhausted form of the resin and the fluid is the eluent at the expected molarity of contact in the strip zone.

Isotherm generation is also accomplished in short-column reactors in which a resin volume is equilibrated with batches of solution that represents a gradation of mixtures of constant normality from pure eluent to pure eluate. This equilibration is obtained by recirculating the fluid through the column for a long time — or a time close to the fluid residence time in the ISEP contact. The recirculating solution is of such large volume that its analysis does not change appreciably. The resin loading is found from a post-equilibration stripping step with a suitable eluent. The complete eluate is collected and analysed which permits calculation of the species concentration on the resin.

A point which should not be overlooked in isotherm development is the effect of time in attaining equilibrium. If the adsorption kinetics are abnormally slow then psuedo-isotherms can be developed that reflect the level of adsorption during a time comparable with that expected in the continuous process. Approach to equilibrium data over a timed period can be useful in process design work but most often the ion-exchange kinetics are fast enough (in the order of seconds and minutes) not to have to worry about this aspect of the process. There are exceptions to this rule though, especially in the area of large-molecule adsorption. The following discussion treats the use of isotherms in the process design phase.

Adsorption isotherms

Fig. 10.10 shows a favourable isotherm in which the adsorbate has a higher affinity for the resin than does the counter-ion being desorbed. Steep adsorption isotherms like this make adsorbate recovery less sensitive to poor stripping performance. In

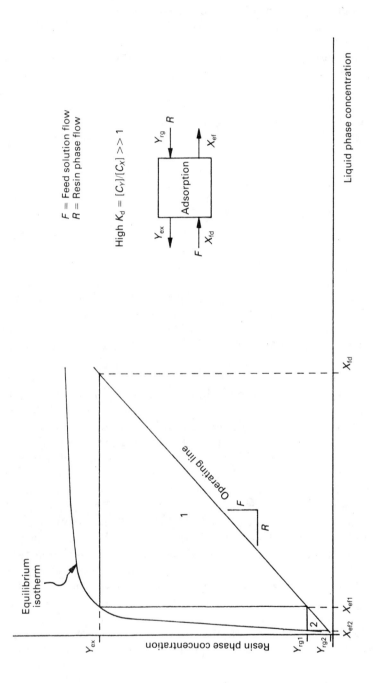

Fig. 10.10 — Adsorption operation: favourable isotherm.

these cases the choice of operating line will influence the resin use. A shallow slope will lead to low loading while a steeper slope will increase the resin loading up to some maximum defined by the isotherm.

In contrast, Fig. 10.11, shows a favourable isotherm with a shallower slope compared to Fig. 10.10. The same performance requires a longer MTZL (more stages). Maximum resin loading becomes more sensitive to adsorbate concentration. Stripping efficiency is most important to adsorption recovery since a residual level of adsorbate left on the resin after stripping now limits the thermodynamically possible effluent adsorbate concentration. With shallow isotherms it is always advisable to consider how wash effluent is reintroduced into the adsorption zone. If introduced with the new feed it dilutes the actual feed concentration and thereby reduces the maximum attainable resin loading. The diluted wash effluent can join the adsorption flow mid-way down the zone so that resin still contacts the most concentrated feed prior to exiting the adsorption process.

Fig. 10.12 illustrates an unfavourable adsorption isotherm (concave downwards). The phase flow ratio choice is critical (at some resin loading) to avoid 'pinching' the isotherm. Investigation of temperature influence on the isotherm slope can often be of use here. Cold temperature isotherms are often more favourable. In such a case the kinetic data must be coupled with the equilibrium data to assess bed-length requirements. As can be seen it is critical to strip the resin to very low $C_{stripped}$ levels so that adsorption effluent specifications can be achieved. Employment of very shallow operating lines is essential in order to achieve low $C_{effluent}$ concentrations. This restricts the maximum resin loading achievable.

Stripping isotherms

The direction of mass transfer is from the resin phase to the liquid phase and therefore the operating lines are placed on the side of the isotherm closest to the phase from which the mass is transferred. In our convention we therefore place the stripping operating lines above the equilibrium curve. Note that the opposite case rules in adsorption.

Fig. 10.13 is shown as a stripping isotherm. It is concave downwards which in stripping is ideal for the same reasons that convex upwards isotherms favour good adsorption conditions. The more concave, the fewer theoretical stages are required and there are no 'pinch points' to worry about.

The unfavourable case is Fig. 10.14 and here the choice of operating line slope is more critical in order to accomplish the required degree of stripping. Use of dual operating lines in stripping, illustrated in Fig. 10.15, helps maximize use of regenerant. The wash effluent can be allocated so that a shallow operating line is employed in the stages of strip closer to the eluent addition point, while a steeper slope is used at the product end of the strip cascade. The optimization of this arrangement is not a trivial task. The case shown (Fig. 10.15) does not take into account that, within a stripping zone in which total solution molarity can change, there can equally be a change in the equilibrium isotherm relationships. This is where accurate modelling techniques can be usefully employed.

The stripping isotherm development must address not just temperature variation but also the effect of eluent strength on the isotherm. The mechanism of stripping

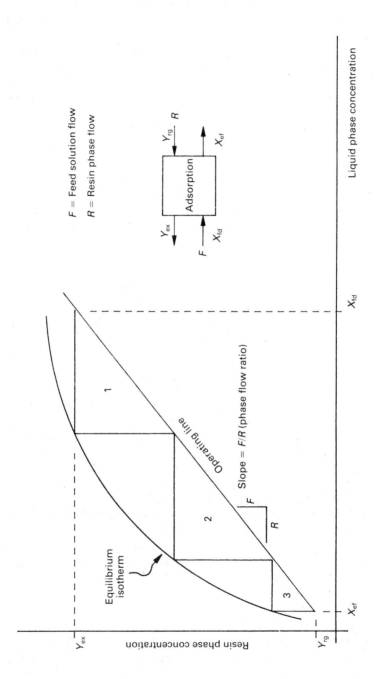

Fig. 10.11 — Adsorption operation: favourable isotherm.

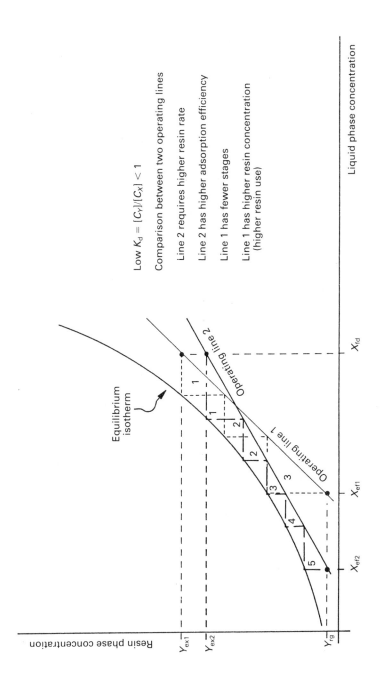

Fig. 10.12 — Adsorption operation: unfavourable isotherm.

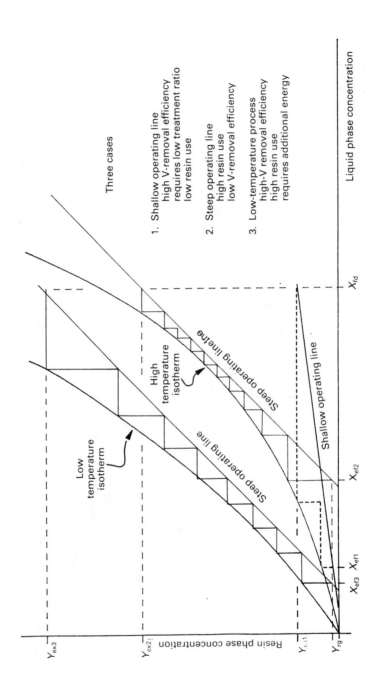

Fig. 10.13 — Vanadium adsorption system.

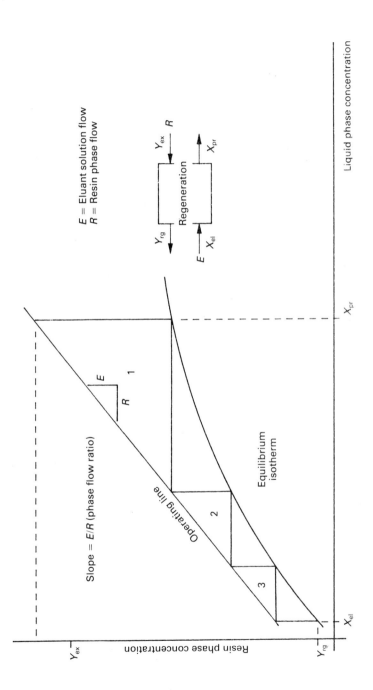

Fig. 10.14 — Regeneration operation.

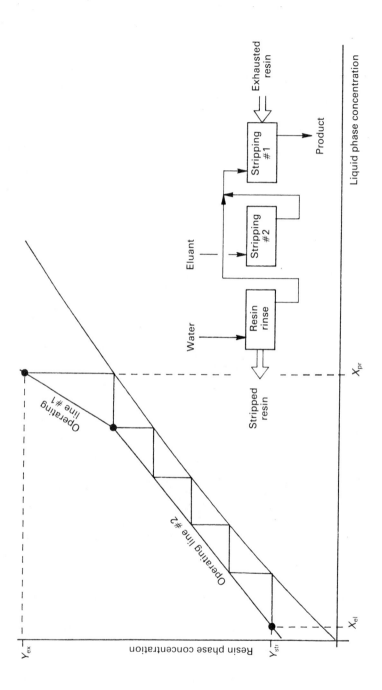

Fig. 10.15 — Stripping isotherm system.

should be carefully evaluated in order to best configure the strip zone. If the stripping kinetics are rapid then the stripping zone can be designed around equilibrium constraints. This means that we need to determine the optimum residence time and eluent strength combination in order to minimize the size of the strip zone.

If the stripping exchange is followed by chemical reaction in solution, making the desorption appear to be irreversible, then we should concern ourselves with minimizing the strip zone solution volume so that product concentration is maximized. This is the case for lactic acid where irreversible stripping occurs provided the strip solution pH remains well below the pK_a of lactic acid. This case is equivalent to having a highly favourable strip isotherm.

It can be seen that isotherm data for the particular adsorption and stripping conditions of the separation provide the designer with a very valuable tool that essentially defines the thermodynamic envelope within which he can operate. The data become exceedingly valuable when the undesirable case of unfavourable isotherms presents itself. The engineer can in such cases choose between low resin use and good adsorption efficiency or high resin loading and poor adsorption recovery.

Isotherm data is collected in the laboratory using unsophisticated equipment. The next section deals with the traditional ion-exchange testing technique from which valuable kinetic information can be obtained to couple with the equilibrium information and permit equipment sizing calculations.

Column testing

Breakthrough and elution tests

The column tests that are traditionally performed to evaluate ion-exchange application are also the basic tools used by AST with certain modifications to customize test parameters best for extracting continuous ISEP system design data.

Initial trials of a sorbent either employ batch or column contacts to determine whether separation occurs quantitatively. A column test is easily done and provides the experimenter with information on kinetics, resin capacity, pressure drop, etc. Much of this information can be usefully used in the conceptual design of the ISEP unit.

A typical column breakthrough profile is shown in Fig. 10.16. The breakthrough profile can provide resin capacity and mass transfer zone length requirements. The influence of isotherm data will effect the number of bed volumes to treat the resin with. In the case of the unfavourable isotherm (Fig. 10.12) it is important to know which is more important, low effluent concentration or high resin loading. The choice influences the feed:resin treatment ratio. A low number of bed volumes is chosen when a low effluent specification is needed.

If the column length is insufficient to contain the mass transfer zone then a second test can be run with a deeper bed to determine the minimum resin bed length. Running adsorption at different bed depths and at the same velocity will provide good information about the stability of the mass transfer front. If the profile shapes are the same (parallel) on exiting both the shorter and the longer column then a stable profile has been established. If not then one can conclude that in the shorter

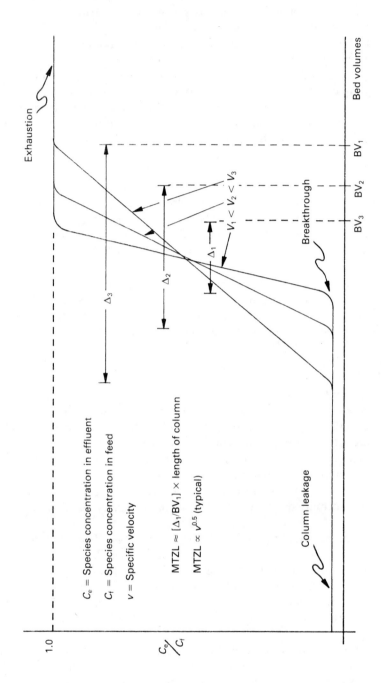

Fig. 10.16 — Typical breakthrough curve.

column the stable profile had not yet been established. Remember that an unfavourable isotherm will typically never result in a stable mass transfer profile. The profile increases in length with time [1]. The resin loading, C_{Rmax} can be calculated from the test mass balance. A confirmation of this loading is calculated from the subsequent resin stripping test.

Adsorption column tests are run at different specific flows (fluid velocity, v) to evaluate its impact on the mass transfer zone length requirements. Often the MTZL α $v^{0.5}$ when the rate-determining step is the mass transfer date. Such a relationship will dictate that we operate at the maximum feasible velocity in order to maximize resin use. The constraints on operating at elevated velocities are: valve operating pressure capability, stripping and washing efficiency consideration. The wash requirements influence the adsorption flow velocities since normally wash zone effluent is returned to the adsorption zone. This is illustrated in the case study examples.

The elution profile is shown for a typical case in Fig. 10.17 and its information again is most useful for designing the continuous system. The choice of elution velocities is influenced by the stripping isotherm and the resin adsorption test. The isotherm will indicate realistic flow ratios in order to accomplish the required degree of resin stripping. It is important to understand that poor stripping is usually the cause of inadequate adsorption. The adsorption test will provide an estimate of the resin exhaustion rate. In a countercurrent system this is a physical flow of solid resin phase. Experience then estimates an approximate chamber diameter and bed depth, which in turn permits the experimenter to estimate the range of practical eluent velocities. These velocities are then used in the elution testing.

For example,

resin flowrate	= R m^{3}h^{-1}
eluate flow	= E m^{3}h^{-1}
[eluate]	= C_{el} g/l,
chamber area	= A m^2,

then specific elution velocity $v_{el} = (E + R \times \varepsilon)/A$ mh^{-1}, where ε = column void volume ≈ 0.65 BV.

Information from the elution profiles confirms the resin loading estimate, C_r^{max}, when the mass balance is calculated.

Elution testing at various velocities at and above the minimum can elucidate the nature of the relationship between elution volume and elution time. Increased elution velocity can have the advantage of operating at lower-resin-residence time in stripping, but suffers the disadvantage of lower eluate concentration and eluent efficiency.

Wash profiles give similar information to that developed in the elution profiles. Residence time information can be extracted in the same manner. Knowledge of the 'number of stages' need to accomplish an adsorption or stripping objective is valuable information in the design of the pilot-unit configuration. The missing piece of information is the HETP. HETP estimates can be obtained from the breakthrough curves by plotting the derivative of the breakthrough profile slope which typically

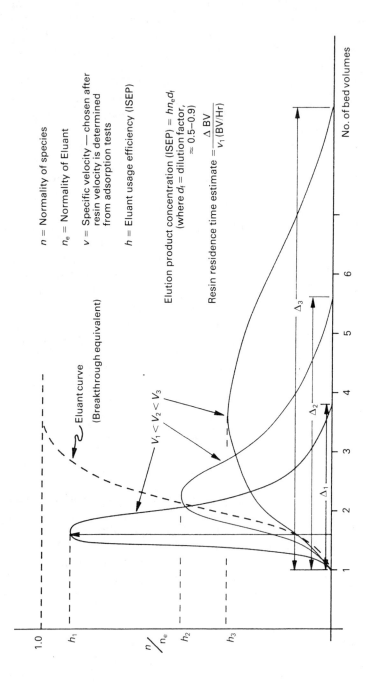

Fig. 10.17 — Typical elution curve.

gives a bell-shaped curve [2] which can then be treated in a manner identical to the chromatographic pulse response curves (refs) as described later in the Chromatography Experimental Procedures section. HETP for elution conditions can be estimated through trial-and-error routines that employ an equilibrium model of a batch column [3]. Alternatively elution HETP estimates can be generated through a pulse test in which the feed solution is loaded and is then followed with the eluent. Taking the derivative of the trailing edge of the elution profile can be used as a last resort to generate a pulse profile from which the HETP estimate can be obtained.

Laboratory simulations
Laboratory simulations of ISEP zones, individually or combined, is practised as an optional but more cost-effective way of checking the performance of a resin liquid system. A typical laboratory set-up is shown in Fig. 10.18. A simulation consists of stepping columns of resin through the same sequence they would experience in an ISEP contact zone with respect to volume of fluid, mass flow of dissolved species, etc. The transfer of fluids through the beds commences with emptying the common effluent containers from the right-hand end of a zone and then passing each fluid batch forward to the right. The feed aliquot to a zone is then passed through the first (most left-hand) column in the zone. At the end of this operation the column containers should all be full again. If one zone's effluent feeds into another zone's feed aliquot then the zone producing the intermediate effluent is processed first. The result of all this is to produce an aliquot of effluent from each product point per step in a cycle. The resin beds are moved one position to the left after all zones have been serviced. A cycle is complete when a column returns to its original position. Usually two to three cycles are sufficient to attain steady state.

Resin/sorbent selection
Resin and sorbent manufacturers have much experience in the application of their sorbents. Plus they know how well sorbents handle specific environments. Some process factors that are important in sorbent evaluation are:

> presence of oxidants and their impact on sorbent activity,
> presence of organic contaminants and their fouling potential,
> evaluation of severe osmotic shock on resin particle size,
> temperature swings and sorbent resistance to heat transfer.

AST normally conducts automated cycle tests on the final sorbent of choice whenever there is a danger of shortened sorbent life. Any activated carbon process employing chemical regeneration is always checked for sorbent poisoning in a 20–50 cycle test. Organic species adsorption processes are cycled to monitor the adsorption capacity, looking for any indication of fouling. Similar testing is done on any sorbent. An interesting observation to illustrate how valuable your own in-house experimentation is was that of not rejecting a resin on the basis that it had poor resistance to osmotic shock. A case in point was the choice of a weak-base resin to serve in the HSO_4^-–lactate^{-1} system. The resin was known to be a poor performer from the

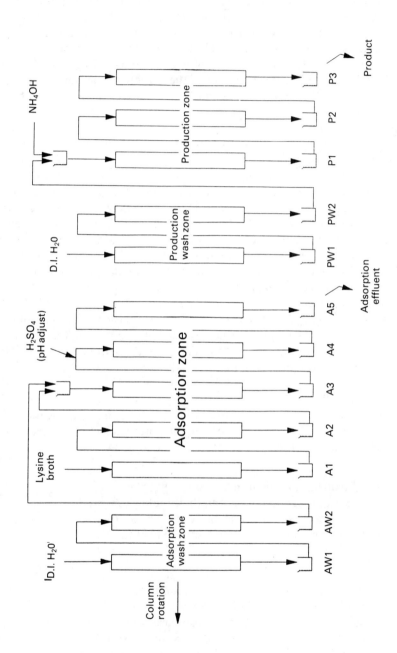

Fig. 10.18 — Benchtop simulation scheme.

osmotic standpoint. Careful initial conversion of the free-base form to the protonated form under controlled conditions with dilute sulphuric acid was all that was required. Subsequent cycling between two ionic forms caused no appreciable resin bead breakage.

The use of an ordinary laboratory microscope often reveals the nature of lost sorbent capacity. Scaling of beads, osmotic breakage and plugging of pores can easily be detected. Fouling due to oil present in an inorganic feed stream would foul the resin within weeks. The solution found here was to add a small amount of non-ionic detergent on a continuous basis to prevent oil adsorption. Another solution was to pretreat the oil-containing feed stream with a batch column of sorbent specially designed to adsorb the organic component.

Table 10.4 is a tabulation of a resin evaluation programme using equilibrium loading capacity as the selection criteria. The application was vanadium adsorption from a phosphoric acid solution. Duolite A-368 was the superior candidate and became even more so after its coarse fractions were screened out. Although in this case the resin selection was based on capacity there was a programme planned later to evaluate the same resins on the basis of resistance to oxidation.

Table 10.4 — Resin evaluation — vanadium adsorption capacity

Resin	Type	Stated capacity		Vanadium use
		Expansion %	mg-equivs/l	mg-equivs/l
MSA-1[1]	Strong base Macroporous	8.4	1000	69
HPQ[2]	Strong base Pyridine	2.6	900	83
A-368[3]	Weak base Macroporous	51.2	1700	85
21K[1]	Strong base Low x-linking	−3.2	1250	36
WGR[1]	Weak base Epoxy-amine	62.1	1140	10
A-368[3,4]	Weak base Macroporous (screened)	49.7	1700	122

[1]Product of Dow Chemical Company.
[2]Product of Reilly Industries.
[3]Product of Rohm & Haas.

A further example of resin selection is shown in Table 10.6. In this case the δP in the adsorption zone assumed special importance owing to the presence of solids in the feed material, lysine fermentation broth. Adsorption capacities and kinetics were comparable for all resins but as can be seen the Bayer resin demonstrated superior pressure-drop performance.

Lactic acid purification from fermentation broth
This new process is a simple anion-exchange technique on a lactate salt feed broth. Lactic acid produced via fermentation is extracted typically from an ammonium lactate liquor. The commercial production of purified lactic acid from fermentation both is usually a process that employs direct acidification followed by ion exclusion, or cation exchange followed by ion exclusion or electrodialysis.

The process employing ISEP continuous ion-exchange uses a strong (or weak) base resin that cycles through the reactions listed below.

Strong base anion reactions

Adsorption	$R_2\text{-}SO_4 + 2Lac^- + 2NH_4^+$	$\rightarrow 2R\text{-}Lac + (NH_4)_2SO_4$
Stripping	$2R\text{-}Lac + 2HSO_4^- + H^+$	$\rightarrow 2R\text{-}HSO_4 + 1HLac$
Washing	$2R\text{-}HSO_4 + H_2O$	$\rightarrow R_{22}\text{-}SO_4 + HSO_4^- + H_3O^+$

Solution reactions

Lactic acid dissociation	$HLac \rightarrow H^+ + Lac^-$	$pK_a = 3.86$
Sulphate/bisulphate	$HSO_4^- \rightarrow SO_4^{2-} + H^+$	$pK_2 = 1.92$

The process relies first on the excellent selectivity for lactate loading in the adsorption zone to displace sulphate/bisulphate ions, then followed by a carefully controlled strip of lactate using sulphuric acid. The pH of the strip zone is kept as high as possible, consistent with full lactate desorption. The desorption of a weak-acid anion with a strong acid is a powerful technique provided there is at least a two point difference in the pK values of the two acids. There is potential to use the technique to separate weak acids of widely differing pK values.

The washing step after adsorption is critical to the purity of the final product. The post-adsorption wash is conducted in countercurrent fashion and removes stray proteins and non-ionic sugars.

Fig. 10.19 is a flow concept of how an ISEP would be configured for this particular separation. The laboratory breakthrough and elution curves are shown in Figs 10.20 and 10.21. These represent two different flow rates and it is from these two curves that the system can be sized. The adsorption test in Fig. 10.20 was run at 12 mh^{-1} velocity while Fig. 10.21 shows results for 7.4 mh^{-1}. Washing and elution flows were the same as for adsorption. 12 mh^{-1} velocity is close to that designed into the commercial unit.

Lactate feed broth enters the adsorption zone flowing into ports 5–7 which are connected together. The complete adsorption zone consists of nine total ports. These nine ports are divided into three groups of three ports. The feed broth passes through the resin bed three times and then exits as an ammonium sulphate by-product stream from ports 11–13. Resin enters the adsorption zone from the right fully regenerated in the sulphate form. As it traverses the zone it contacts an ever-richer lactate solution and loads to its equilibrium level with the feed broth. This equilibrium loading can be greater than 1 meq/ml. The fluid following in countercurrent manner

Ch. 10] Continuous adsorption and chromatography 193

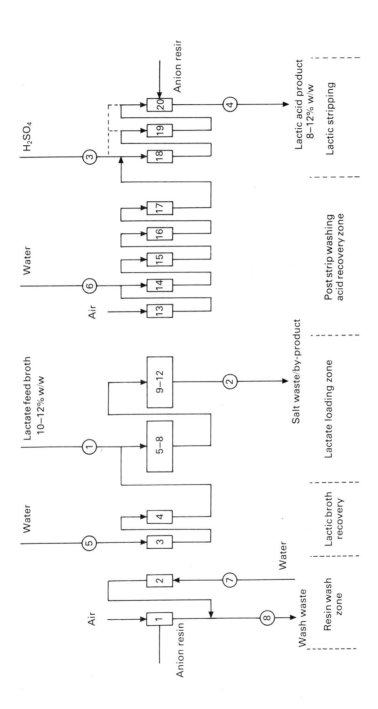

Fig. 10.19 — Lactic acid purification.

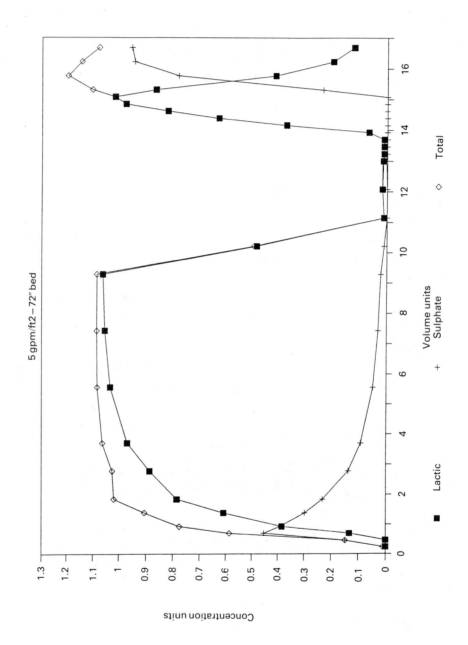

Fig. 10.20 — Lactic acid BSE-1.

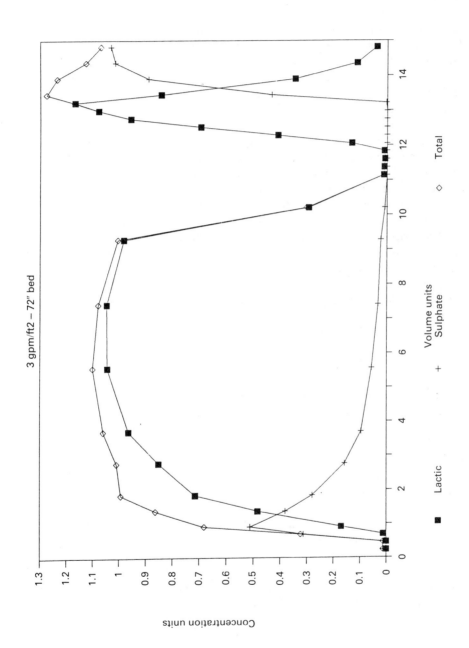

Fig. 10.21 — Lactic acid BSE-2.

passes through three lengths of bed in order to fully transfer the lactate species onto the resin.

The resin exiting the adsorption zone is first washed in a three-stage countercurrent water adsorption wash zone (ports 2–4) and is then drained with an air assist (port 1). The adsorption wash-zone effluent containing entrained lactate feed broth and some water is then re-introduced to the adsorption zone in the second pass at ports 8–10. This permits the resin to contact the highest feed concentration and yet still process the wash-zone effluent through the adsorption zone. The drain effluent can be reused as wash water feed but is often dumped to the sewer to avoid the use of an extra pump.

Drained resin now enters the lactate strip zone where a two-pass countercurrent arrangement (ports 19 and 20) is employed for the main fluid flow. Acid is added into each pass of this two-pass system to maintain an effluent pH at the desired level between 1.5 and 2.5. The stripping process is essentially irreversible and its kinetics are rapid. Maintaining a pH as high as 2 in the product means that the stripping agent (sulphuric acid) is present at very low concentrations in the final product.

The stripped resin flowing out of the strip zone is washed free of sulphuric acid which is returned directly to the strip zone. The wash zone is a four-pass series countercurrent system (ports 15–18). The resin itself, once washed, is ready to re-enter the adsorption zone. The washing process also converts the HSO_4^{1-} form resin into a SO_4^{2-} form, considered more suitable for the adsorption reaction. A post-wash drain (port 14) is incoprporated in order to minimize the water in the ammonium sulphate stream which might go to evaporation.

Experimental work
The feasibility of the process was demonstrated in early laboratory breakthrough work which showed quantitative lactate loading on anion resins and easy stripping with sulphate-free lactic acid fractions in the effluent. The breakthroughs in Figs 10.20 and 10.21 are interesting in the ease of loading and desorption that is indicated. The kinetics of the operation are excellent as evidenced by the steep adsorption and elution breakthrough slopes. The fast elution requires only 2 BVs at nearly 7BV/hr flow rate requiring therefore about 15–18 minutes in stripping. Washing is equally rapid. Adsorption is given a little more than the 2 m bed depth to better eliminate slight leakage.

The lactic acid product purity is paramount and therefore demanded that we pay more than the normal attention to post-adsorption washing. Broths contain sugars, amino acids, proteins and other organic acids as impurities. Poor resin washing will permit these species to contaminate the product. The results of the washing studies were based on colour generation in the lactic product after a heat-stability test. Any carbonizable organic species will burn and discolour the product if present. The major findings were:

(1) washing improves with increasing fluid velocity,
(2) washing is much improved with increasing resin time in the wash zone,
(3) residence time is more influential than fluid volume employed.

Washing after lactic stripping was also studied to investigate the behaviour of sulphuric acid regenerated resin in a water wash zone. Much water is required to convert all the HSO_3 to SO_4 at least at ambient temperatures with D.I. water. Further study of this wash is merited.

Lysine fermentation broth purification

The flowsheet for the process is shown in Fig. 10.22 and illustrates many of the truly flexible characteristics of the ISEP unit which are commented on below. Lysine purification via ion exchange is not unknown in industry but has meant the design of very large batch columns. It is only because of the high value of the product that such extravagant designs are tolerated. An example of the difference between batch and continuous designs has already been discussed in the Adsorption Process Design Procedures section. The ion-exchange basis for the purification of lysine broth is described by the following two equations.

Strong acid cation-exchange reactions

Adsorption $\quad 2R\text{-}NH_4 + HLysH_2^{2+} \rightarrow R_2\text{-}HLysH_2$
Stripping $\quad R_2\text{-}HLysH_2^+ + 2NH_4^+ + 2OH^- \rightarrow 2R\text{-}NH_4 + HLys + H_2O$

The process basis exploits the dissociation equilibrium chemistry of lysine in acidic and basic environments. Thesee dissociations are listed below and are illustrated graphically in Fig. 10.22.

Lysine dissociation equilibria (see Fig. 10.23)

$$NH_2-C\begin{array}{c}NH_2\\|\\-C\\|\\COO^-\end{array} \quad pH>10 \quad -1 \quad pK_\varepsilon = 10.53$$

$$NH_3^+-C\begin{array}{c}NH_2\\|\\-C\\|\\COO^-\end{array} \quad 7<pH<10 \quad 0 \quad pK_\alpha = 8.95$$

$$NH_3^+-C\begin{array}{c}NH_3^+\\|\\-C\\|\\COO^-\end{array} \quad 2<pH<7 \quad +1 \quad pK_1 = 2.18$$

$$NH_3^+ + C\begin{array}{c}NH_3^+\\|\\-C\\|\\CLOOH\end{array} \quad pH<2 \quad +2$$

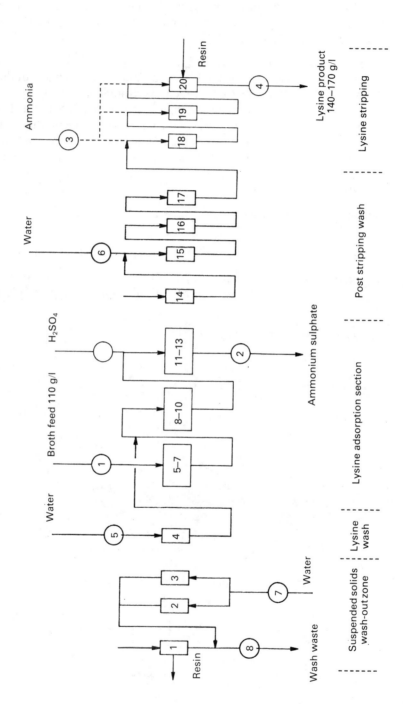

Fig. 10.22 — Lysine purification.

Ch. 10]   **Continuous adsorption and chromatography**   199

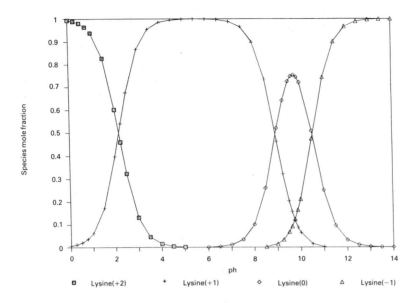

Fig. 10.23 — L-(+)Lysine dissociation equilibria. Species distribution vs. pH.

The affinity for the strong acid resin is high for the +2 species which has a steep isotherm, whereas the +1, mono-protonated species has a much shallower isotherm. Fig. 10.24 shows how, from an equilibrium standpoint, we approach the

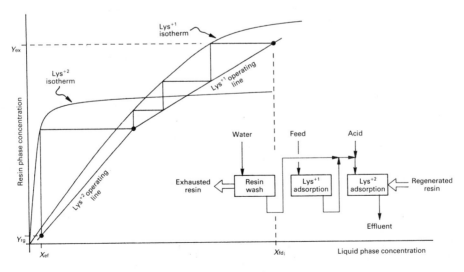

Fig. 10.24 — Lysine adsorption system.

adsorption zone design in order to maximize the resin loading for lysine.HCl while achieving high lysine recovery.

Lysine broth feeds assays 70 to 130 g/l lysine.HCl. Protonated lysine is adsorbed in exchange for ammonium ion in the ISEP adsorption zone. Adsorption-zone liquid effluent contains ammonium sulphate. Lysine-loaded resin passes into the post-adsorption wash zone. The fermentation broth is fed to the ISEP *unfiltered*. Prefiltration is a prerequisite for fixed-bed ion-exchange systems owing to the length of the beds employed. In the ISEP almost all the cell mass passes through the shallow beds into the adsorption-zone effluent. The remainder is washed out in the upflow wash zone. Pre-acidification is unnecessary when the broth comes directly from fermentation. Acid addition is made at the effluent end of the adsorption zone. This *mid-zone pH adjustment* is vital to good lysine recovery and high resin loading. Pilot-plant runs demonstrated clearly that mid-zone acid addition can increase recovery by up to 10% — Table 10.5 compares the data from pilot runs with and without mid-zone acidification. Modification of the solution chemistry in mid-zone is a unique capability of the ISEP contactor.

Fig. 10.24 shows the concept of how we can take advantage of the strong affinity of Lys^{+2} to load the resin efficiently using two sites per lysine molecule but ensuring a low lysine residual in the adsorption effluent, C_{ef}. Note how on the steep isotherm the lysine residual in solution is less sensitive to incomplete stripping. Operating line #2 employs the Lys^{+2} isotherm for the reaction,

$$2R\text{-}NH_4 + HLysH_2^{+2} \rightarrow R_2HLysH_2 + 2NH_4^+$$

Once the resin is loaded with Lys^{+2} it then enters the zone of higher pH, with increased $[Lys^{+1}]$, which has *not been diluted with wash water effluent* and therefore has the potential to compete for sites with Lys^{+2}. Loading is maximized when the resin loading reaches that value consistent with resin in equilibrium with fermentation broth feed solution. The process operator needs to be prepared to increase his resin flow rate if recovery is depressed. The operating line #1 shown in Fig. 10.24 is for the monoprotonated lysine loading reaction,

$$R_2\text{-}HLysH_2 + {}^-LysH_2^{+2} \rightarrow 2R\text{-}LysH_2 + H^+$$

The exact mechanism for switching sites from Lys^{+2} occupancy to Lys^+ residence is not known at this time. The observed facts from pilot data are that high lysine recovery cannot be gained simply by adding the acid to the adsorption feed. Acid injection downstream and wash effluent introduction to prevent feed dilution influence the ultimate lysine loading on the resin and the lysine recovery in a given fixed length of bed. Table 10.6 provides pilot data for comparative adsorption conditions.

The lysine stripping consists of converting the lysine-protonated species to a neutral uncharged form that has no further affinity for the resin sites. This is accomplished by holding the pH at a level around pH = 9.5 to 10.0. There will always be a tendency for the resin Lys^{+2} species to desorb and equilibrate with the small amount of NH_4^+ which exchanges onto the resin sites. This reaction is not a very

Table 10.5 — Lysine purification

General	
Bed depth	0.91 m each cell
Resin type	Dowex strong acid cation
Rotation rate	1.0 hr
Adsorption zone	
New feed flow	1.14 BVhr^{-1} unfiltered fermentation broth
Feed concentration	70–130 g/l lysine.HCl
Adsorption feed to pass #2	2.15 BVhr^{-1}
# Passes	3
# Ports/pass	3 Acid added to pass #3
Specific flow	10.5 mhr^{-1}
Zone effluent	1.78 BVhr^{-1}
Post-adsorption wash zone	
New water feed flow	1.00 BVhr^{-1}
# Passes	2
# Ports/pass	1
Specific flow	14.7 mhr^{-1}
Zone effluent	1.00 BVhr^{-1} to adsorption zone pass #2
Backwash zone	
Water feed flow	2.00 BVhr^{-1}
# Passes	1
# Ports	2 Upflow + (air (one port only)
Specific flow	14.7 mhr^{-1}
Zone effluent flow	2.39 BVhr^{-1} includes post zone drain flow
Post zone drain	yes
Drain return	no
Production zone	
Ammonia feed	2.82 kg-eqm^{-3}
Ammonia concentration	3–4 Normal
# Passes	3
# Ports/pass	1
Specific flow	14.45 mhr^{-1}
Zone effluent	0.60 BVhr^{-1}
Post-production wash zone	
New water feed flow	0.60 BVhr^{-1}
# Passes	3
# Ports/pass	1
Specific flow	14.45 mhr^{-1}
Zone effluent	1.00 BVhr^{-1} to production zone feed tank

The process design chronology followed the steps; laboratory column tests, resin evaluations, bench-scale simulation, pilot-plant trials, commercial operation.

Table 10.6 — Resin evaluation — δP for lysine broth

Resin	0 hours	0.5 hours	1 hours	2 hours	3 hours	4 hours	8 hours
Dow/Dowex XUS 40197	3.0	3.0	3.0	3.5	4.0	4.0	4.0
Rohm & Haas/IRA 120+	6.0	7.0	7.5	7.5	7.5	7.5	7.5
Mitsubishi/Diaion SKL 10	0.5	2.5	3.0	3.0	3.0	3.0	3.5
Bayer/Katalysator	0.5	1.0	1.0	1.0	1.5	1.5	1.5
Rohm & Hass/XE-636	2.0	3.0	3.0	4.0	4.0	4.5	4.5

Testing in a 36″ long resin bed (BV = 180 ml) operating at 5 gpm/ft^2 (12.2 mh^{-1}) with a recirculating broth volume of 360 ml.

promising exchange except for the accompanying neutralization reaction that converts Lys^{+2} to Lys$_{uncharged}$. This causes NH_4^+ to form as a counter-ion and load onto the resin. The resin conversion to the NH_4^+ form occurs because the reaction is driven to completion provided the pH is maintained at around 10.

The post-adsorption wash is an important step in that it must recover entrained lysine broth and return it to the adsorption zone without excessive dilution. The accumulated solids that did not pass through the bed are washed out in this zone using an upflow wash in an expanded bed. Air is injected into the upflow so that a good scouring can occur that prevents solids from being physically attached to resin particles. The wash is divided into two steps; first there is a downflow displacement was that effectively recovers over 90% of the entrained adsorption-zone feed fluid. Port #4 in Fig. 10.22 is allocated to this function; secondly a two-port parallel upflow zone is dedicated to washing the bed clean of entrapped solids. Ports 2 and 3 are used for this purpose. Two ports are necessary in order that sufficient bed volumes can be processed while still using a velocity of upflowing water compatible with an expanded fluidized bed. Cells are designed with 25% to 50% freeboard for this purpose. If solids are not completely removed they cause the stripping-zone lysine product to become cloudy, an objectionable condition.

The desorption zone is a counter-current four-pass zone with a single port for each pass. This configuration has not been optimized and may be switched for a two-pass zone with two ports/pass. Alternatively, if resin time determines the stripping efficiency then a single pass of four or even five ports with slow fluid velocity may be more effective. All these options are available even after the commercial ISEP is in the field since changing the configuration is simply a relatively minor piping modification. The choice of desorption-zone configuration depends on the process downstream constraints, the desorption mechanism and the tolerance to excess eluent available. In the case of lysine, ammonia recovery via air stripping can be practiced and thereby eliminate the constraint of not having excess eluent in the lysine product stream. Assuming that the lysine desorption mechanism is not equilibrium constrained and is a kinetic (diffusion)-controlled process, a single-pass system makes sense where the resin has enough time exposed to the high pH eluent whose net volume now becomes simply a function of the wash-water bleed into the product stream from the two sources; first the water associated with the resin entering the adsorption zone, and secondly from the net excess water in the strip-wash effluent. So the design of the desorption zone now comes down to providing sufficient residence time for the lysine to completely desorb, and minimizing the net volume of water used for product stream make-up.

The use of anhydrous ammonia helps, but only transfers the water associated with recycled ammonia elsewhere. Excess ammonia in the discharge is therefore not without its consequences and should be held to its lowest level. The use of displacement entrainment rejection on the resin entering the strip zone is therefore encouraged.

Experimental work
The column-breakthrough testing results are summarized in Figs 10.25 and 10.26 for the adsorption and eluent profiles. The general performance of all strong-acid resins

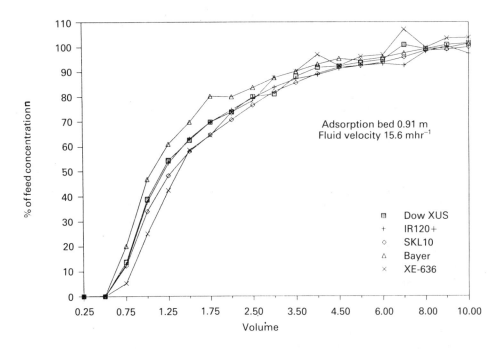

Fig. 10.25 — Lysine purification: adsorption breakthrough profiles.

was comparable in both adsorption and stripping with ammonia. Examination of the graphs indicates that at fluid velocities of $12\,\text{mh}^{-1}$, used in the testwork, the

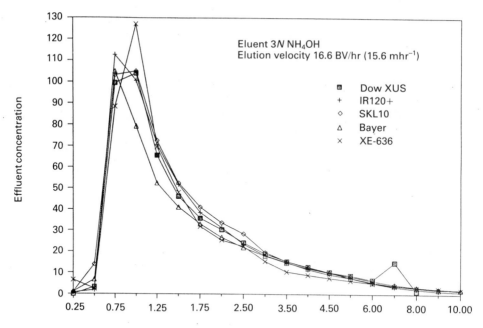

Fig. 10.26 — Lysine purification: elution profiles.

adsorptions and elutions were complete after 7–8 BVs. The flow rate used was equivalent to 16.5 BVh^{-1} which would suggest that the resin be permitted about 25–30 minutes of residence time in the two exchange zones. This would be the first estimate of zone sizing. The breakthrough profile for feed solutions as concentrated as lysine does not provide a firm basis for MTZL calculations owing to the immediate leakage that occurs. Without extensive testing at increasing bed depths it is not possible to determine whether the effluent profile represents a stable bed profile. MTZL estimates from column data only apply to continuous systems that operate at the resin saturation level. In ISEP operation we may choose to operate at significantly lower loadings in order to achieve the desired recovery in less stages. This concept was discussed in the previous section on system design.

Bench-scale simulation was designed based on industry experience. The simulation consisted of rotating 12 columns through the ISEP process sequence in a stepwise fashion (Fig. 10.27). Each step consisted of processing a small volume of fluid through each of the columns. The appropriate water, feed, acid and eluent additions were made in each step. These quantities were set for each simulation; for instance a fixed amount of lysine feed broth/volume of resin (in a column) was a fundamental variable that aimed at using the resin capacity at some desired level; the simulation results would determine how successful the attempt was. Each step produces an aliquot of effluent from each zone; Product (P3) and Adsorption Effluent (A5). These samples are analysed and plotted to view the trend. At the end of each step the columns were rotated one position to the left. A cycle is complete when a column finishes a complete rotation. Up to three cycles are typically needed

Fig. 10.27 — Benchtop simulation scheme.

to come to a complete steady state condition (Fig. 10.28). This is detected by examining the effluent sample values for repeated steady values. At that point the experiment can be stopped and the intermediate solution (e.g. A1–A4, PW1 and PW2, etc.) sampled and analysed to provide a circuit profile for the analysed species. Fig. 10.29 shows such a profile.

Pilot-plant runs followed the basic configuration used in the simulation but employed 0.6 m deep beds. The pilot configuration is illustrated in Fig. 10.30. The pilot runs did provide valuable input to the optimization process of maximizing recovery and lysine-product concentration but in actuality the highlight of the pilot campaign was a successful solution to handling the unfiltered broth. The final configuration is shown in Fig. 10.22. The number of ports required for the backwash function was determined through pilot trials. This is a prime example of the exceptional flexibility built into the ISEP concept.

Resin screening tests for this process were implemented to assess the tolerance to solids plugging of the various candidate resins. All resins had been previously screened as having acceptable performance in adsorption and regeneration. The test procedure consisted of recirculating a fixed amount of feed broth through the resin bed for eight hours while monitoring the pressure drop. The test results are summarized in Table 10.6.

206   Continuous adsorption and chromatography   [Ch. 10

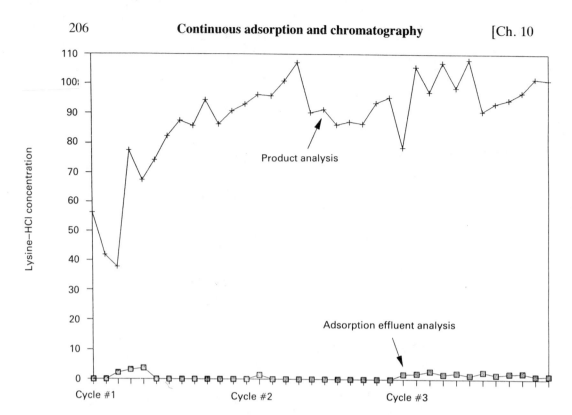

Fig. 10.28 — Lysine purification simulation: effluent profiles.

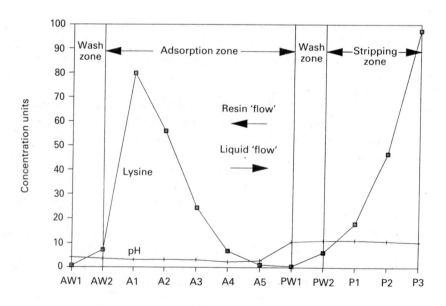

Fig. 10.29 — Lysine laboratory ISEP simulation: column profile.

Ch. 10] Continuous adsorption and chromatography 207

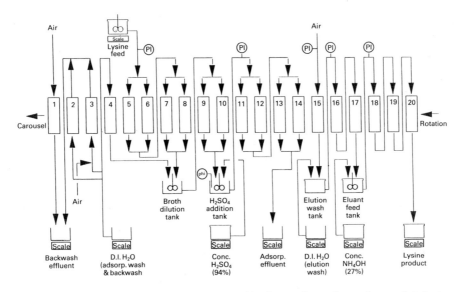

Fig. 10.30 — Amino acid (lysine) recovery and purification continuous ion-exchange pilot plant configuration.

ISEP AS A CHROMATOGRAPHIC SEPARATOR

It probably has not gone unnoticed that the ISEP contactor bears a physical resemblance to the experimental contactor extensively described in the many works of Barker *et al.* Much of that work dealt with fractionation of sugar isomers. The same principle employed by Barker has been successfully commercialized by UOP as 'Sorbex' and IWT as 'ADSEP'. In addition there are at least two commercial systems that employ fixed beds sequenced by valving to simulate a moving bed. The Sorbex and ADSEP systems are used extensively as chromatographic fractionators. They consist of sections of short bed lengths mounted vertically. Such systems are described fully in the literature. There is a fluid-flow down through each section and a recycle-flow back to the top section. This arrangement makes it difficult to adapt the systems to general-purpose adsorption applications similar to those already described for the ISEP.

The ISEP can, on the other hand, be adapted easily to the chromatography mode in either of two ways; first it can be operated as a multicomponent fractionator in which it automates batch column conventional chromatography; secondly it can be configured to mimic the existing technology of Sorbex and ADSEP.

Continuous co-current chromatography

The ISEP contactor is ideal for automating batch chromatography while eliminating much of the operational complexity associated with batch column operations. Fig. 10.31 is a typical configuration for such an automated system.

The batch system sequence is 'programmed' into the fixed ports of the ISEP valve instead of sequencing the introduction of fluids to a single column via many different

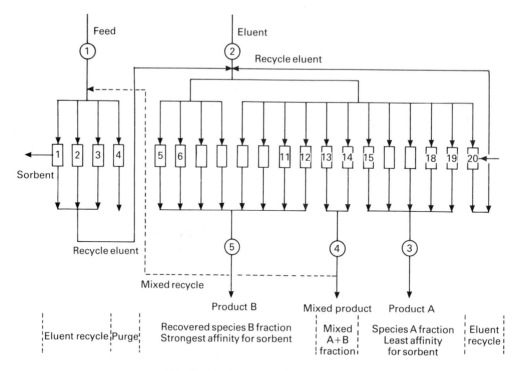

Fig. 10.31 — Continuous chromatography.

valves. Note that in this example the feed pulse is spread over four ports (4/20 = 20% of the time) while eluent is fed over the remainder of the fixed ports. Note that in this example the eluent is manifolded into two major sections, ports 11–20 and ports 5–8. This permits us to independently vary the fluid velocity in the two zones. The higher velocity is employed in the zone for ports 5–8 to accelerate the elution of the pure species B which is the only remaining component in the bed.

The ISEP discharges are manifolded in a manner such that products are removed at desired purities while mixed, off-specification fractions are recycled. Note that, in contrast to adsorption processes, the upper-valve and lower-valve configurations have totally different arrangements. The continuous recycle allows the process to produce only those products desired. The recycle does mean that the system size is increased by the amount of recycle.

You can see several advantages of the ISEP for continuous chromatography through study of the diagram. There is great flexibility in the ability to isolate use of variable fluid velocity. Elution becomes even more flexible in that different strength eluents can easily be accommodated. In the separation of rare-earth fractions, for instance, we would operate at low molarity for initial development of the separation and then increase the molarity after certain species were recovered. Variation of pH becomes a manageable tool in the elution sequence. This has great importance in the separation of amino-acids and organic weak acids.

An example of equilibrium-limited strip is the desorption of $CdCl_3^-$ from an anion resin using water. The relationship of strip residence time and strip flow rate indicates that there is some maximum $[C_d]$ that can be achieved that corresponds to a minimum residence time of the fluid phase in contact with the resin. A plot of % *elution* vs. *time*, for a fixed and sufficient volume and strength of eluent, rises from 0% (t_0) to 100% (t_{100}) eluted at some time and then stays flat indicating that for this bed a slower flow rate (longer residence time). Another plot would be strip flow rate (mh^{-1}) vs. BVs required for 100% elution. This graph should initially be flat in the low specific flow rate range and then rise at some rate proportional to the overall mass transfer kinetics.

The strip zone is shown as a series countercurrent contact with water. The length of the contact is a function of the required resin residence time. Complete resin stripping is a critical aspect of the process. Even small residual cadmium levels on the resin could cause a problem in meeting the adsorption effluent specification.

Batch tanks for recycle mixed fractions are eliminated and recycles become continuous flows often returning to combine with the ISEP feed. This mode of operation makes the ISEP larger but provides a continuous system without by-product manufacture.

The ISEP valve alignment can be adjusted in order to minimize the problem of mixing of effluents on the lower valve that occurs when one chamber is exiting a port while the other is entering — Fig. 10.3. Alignment adjustment is the best way to minimize effluent mixing in a multi-purpose pilot unit. In the case of a commercial machine, the lower valve ports are specially sized smaller so as to optimize the valve geometry for the specific separation.

Binary fractionation or continuous countercurrent chromatography
A more efficient way of employing the sorbent charge is to use its capacity through the separation zone. In the batch column approach only some 15–20% of the sorbent is typically working at any point in time. If, on the other hand, a sorbent/fluid system is designed to emulate a liquid/liquid extraction cascade then we can involve a much higher percentage of the sorbent in the working mode.

Fig. 10.32 is a flow diagram that shows a conceptual ISEP fractionation process in which a feed stream, containing two species A and B, is treated in such a manner that species A, which has a higher affinity for the sorbent phase than species B, is concentrated to a high level of purity, in the sorbent phase and then recovered through an elution step shown on the same diagram. Species B also is purified to a great extent and reports to the liquid effluent stream of the fractionation zone as shown. The species A and B can equally be two chosen fractions into which a multi-component mixture can be split.

The basis for such a separation uses the same principles employed in distillation and liquid–liquid extraction fractionators where two species are distributed between two phases. The difference in the relative affinities for the two phases, the K_{Di}s, is used to displace one component up or to the left (A enrichment section) and the other component down or to the right (A stripping section).

Another visualization of the process is to consider the behaviour of the species in the batch column pulse tests, as shown in Fig. 10.33. Species B exits first, having less

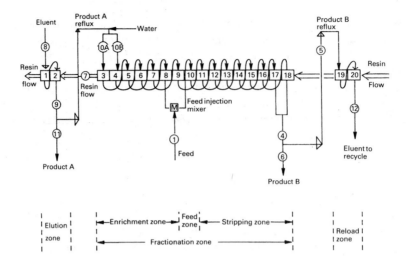

Fig. 10.32 — ISEP binary fractionation.

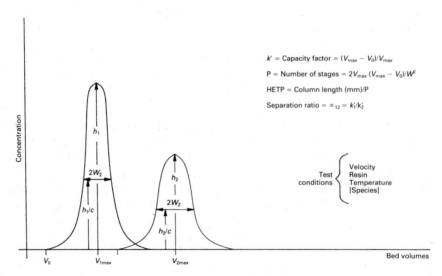

Fig. 10.33 — Batch chromatography.

affinity for the sorbent phase and therefore exhibits a higher velocity through the column. Species A which interacts more strongly with the sorbent is more retarded in its passage through the sorbent bed and exits after species B. This performance difference can be described by assigning two different characteristic velocities to the components; $v_A < v_B$. Velocities v_A and v_B are the rates of migration of A and B through the sorbent bed. Now, if one considers imposing an opposing velocity on the sorbent phase v_{resin} where, $v_A < v_{resin} < v_B < v_{fluid}$ then, we should be able to separate the species into two fractions each richer than the feed in one of the two components. The liquid flow rate v_{fluid}, represents the highest velocity in the system. The extent of the enrichment and recovery obtained is then a function of the difficulty of separation, α, the length of bed in fractionation, the HETP, and the efficiency of the continuous contactor with its interstage fluid transfer and dead volumes, etc.

The features of a true counter-current cascade contactor are illustrated in Fig. 10.34 where the fluid phase L travels through a moving sorbent bed at v_{fluid} and the sorbent phase R travels at v_{resin}. The feed fluid is injected into the cascade at some point F situated between the ends of the cascade. As the feed fluid enters it mixes immediately with the downward-flowing liquid. Note that v_{fluid} has two values; the eluent flow, $L = E$ in the enrichment section above the feed point, F_d, and $L = E + F$, the sum of the eluent and feed liquid flows below the feed point. The ISEP attempts to emulate this ideal solid/liquid two-phase contactor.

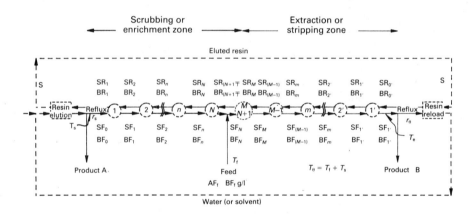

Fig. 10.34 — Binary fractionation stage concentrations.

The moving bed is divided into multiple discrete lengths of sorbent. The liquid exiting the bed is routed to the top of the next bed section via the ISEP valving system. Sections of bed remain in a given position for a fixed length of time. This time varies with each process and is related to the average sorbent flow rate. For example, if each section of bed contained $1\,m^3$ of sorbent, 0.7 BV void volume and the sorbent flow out of and into fractionation where $10\,m^3h^{-1}$ then the residence time in each active port position would average.

$$60\,\text{min}\,h^{-1}/[10\,m^3h^{-1}\,(\text{specific velocity})/1\,m^3\,(\text{void volume})] = 60/[10/0.7] = 4.2\,\text{min}$$

This means that every time the bed moves one section to the left in Fig. 10.32 there are 4.2 minutes of 'misplaced feed' solution to be sent downstream in order to avoid carry over or 'barberpoling' of solution to the left and ultimately contaminating the product A.

The ideal moving-bed contactor attempts to move the sorbent bed continuously against the counterflow of fluid. To date such a system has not materialized. Such a system does not suffer from dead volume of solution external to the bed which occurs whenever a bed section is moved or simulated to have moved. Handling this dead volume is critical to the separation efficiency.

The ISEP valve has been specially designed to work in these situations. Instead of the concept of an upper and lower valve we have combined the two valves onto one so that the piping required to connect upper and lower valve ports in a series sequence has now been reduced to a fraction of the length needed in the standard design.

Some advantages that the ISEP has over other moving bed or simulated moving bed devices are as follows.

(1) Low dead volume < 1% of bed volume
Dead volume is less than 0.3% of the total resin bed volume. This assumes a certain piping arrangement and piping diameter. The amount of dead volume that is associated with 'misplaced fluid' is estimated at one third of this or 0.1%. At this level we do not require elaborate flushing sequences when columns switch ports.

(2) High specific flow rates used
Flow rate = $10-20\,\mathrm{mhr}^{-1}$, shallow beds easily serviced with booster pumps if required.

(3) Higher recovery/unit volume of resin
The result of higher specific flows.

(4) Isolated elution zones
Optional use of chemically different/more efficient eluent.

(5) ISEP valves
ISEP valves eliminate the multitude of solenoid valves. Logic incorporated into valve design — no complex programming station.

(6) Eluent consumption
Eluent usage potentially decreased owing to stripping flexibility.

Modelling fractionation
The theory of fractional extraction developed for liquid systems is directly applicable to liquid solid systems in which the resin phase can substitute for the light liquid phase. The interpretation of the isotherms of the mixed species across the spectrum from pure A to pure B can conceptually be illustrated on McCabe-Thiele diagrams [5]. Examples of these diagrams are shown in Figs 10.35 and 10.36. Fig. 10.35 has the

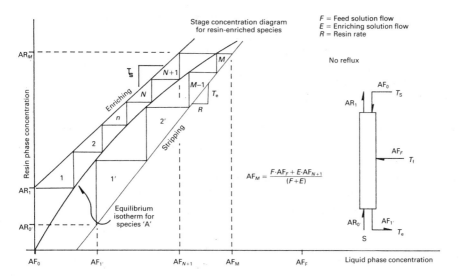

Fig. 10.35 — ISEP binary fractionation.

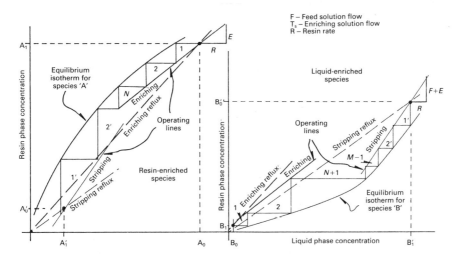

Fig. 10.36 — ISEP binary fractionation isotherm with reflux.

fractionation process plotted for species A while Fig. 10.36 is the diagram for species B. Note that operating line slopes in enrichment and stripping are the same for each species, reflecting the same cascade operating parameters. Reflux lines are indicated and illustate how the same separation can be effected while decreasing the number of theoretical stages, increasing the product concentrations but at the cost of higher

$L:R$, liquid-to-resin-flow ratios. The economic impact of reflux needs evaluation on a case-by-case basis.

Certain equilibrium relationships permit the use of reflux to actually increase the product concentrations to values above the feed values. This relationship is illustrated in Fig. 10.36 where operating lines are located between the two equilibrium lines. This situation can only be practically accomplished through the use of reflux [6].

The important observation to make with respect to the equilibrium data development work is to appreciate the limitations imposed by the resin/sorbent maximum capacity for the various pure and mixture species as a function of %DS (dry solids). There is usually great advantage in operating at as high a feed and cascade concentration as possible since this minimizes the liquid flow rates in the two separation zones. ISEP size and cost are sensitive to liquid flow rates.

The equilibrium lines also delineate the various possible regions of phase-flow ratios that can be used, including some insight into the reflux ratio limitations and attainable product purities. All of the graphical techniques can, in principle, be modelled. A simple model employs stage-by-stage calculations and linear constant-equilibrium distribution coefficients. Initial conceptual sizing is accomplished using such a model. For severely non-linear equilibria, the separation can be segmented into sections of different values for the distribution coefficients, an approach to continuously varying equilibrium values.

The simple model used to construct the flowsheet for continuous fractionation is described by the following equations and is pictured in Fig. 10.34.

$$X_M = X_1 \frac{[S_E^M - 1 - r_E(S_E^{M-1} - 1)]}{(S_E - 1)y} \quad (1)$$

$$y_{n-1} = y_1 \frac{[1 - S_S^{N+1} - r^S S_S(1 - S_S^N)]}{S_S^N (1 - S_S)} \quad (2)$$

S_E = $K_i.S_E/(T_F + T_S)$
S_S = $K_i.S_S/T_S$
S_E = $S_S = S$, resin flow rate m³h^{-1}
i = components A and B to be fractionated
K_i = equilibrium phase distribution coefficients, $[C_{resin}]/[C_{liquid}]$
y_{N-1} = $K_i.X_M$, for each species at the feed stage
y^1 = $x_0.T_S/(r_S.S)$, enrichment or scrub reflux line through origin
y_0 = $x_1.r_E.T_E/S$, stripping or extract reflux line through the origin.

As can be seen, the major assumption of this model is the constant value for the K_s relationships which typically are variable as a function of total solution concentration.

The model input consists of:

(a) the flows, T_S, T_F, S, m³hr^{-1}
(b) feed mass flows and concentrations, A,B kghr^{-1} and A,B kgm^{-3}

(c) reflux fractions (r_E, r_S),
(d) distribution coefficients, K_A and K_B and,
(e) the recovery and purity of one component, R_A and f.

The mass balance around the system is first calculated based on the recovery, purity, flow rate and reflux specifications. With the mass balance and effluent concentrations defined, Eqs (1) and (2) are manipulated to convergence at the feed stage simultaneously for both species. The number of enrichment (N) and stripping (M) stages are varied to accomplish the convergence. After convergence of a particular step the results must be checked to see that no solution or sorbent concentrations have exceeded the allowed maxima from experimental data. The effluent streams must also be compatible with production of the required reflux stream strength unless the operation plans to include an external means of concentrating the reflux system.

All this preliminary analysis permits the experimenter to set up a reasonable separation on the continuous ISEP machine through use of the knowledge of the HETP from pulse tests and N, the number of stages for a given purity and recovery.

Once the column is stabilized and operating well then the experimental programme commenced to explore the response of the system with a view to optimizing the sorbent productivity. Sorbent productivities are typically expressed as:

Productivity — feed basis = [kg-feed.hour^{-1}].[metre^3 sorbent],
Productivity — product basis = [kg-prod.hour^{-1}].[metre^3 sorbent],
@ specified *purity* and *recovery*,

Productivity data are the best initial measurements to employ in order to assess the size and cost of a commercial system. The system parameters to vary during the testwork are listed here,

Feed pulse size is directly related to productivity on a feed basis but must be optimized with respect to recovery at some desired purity. Too large a pulse will eventually overload the column. When a fully conditioned bed receives a feed pulse the species to be separated distribute themselves between the resin and liquid phases at the top of the column up to the resin-holding capacity. If the resulting pulse spreads too far down the column there will not be enough effective column length in which to develop the separation.

When the separated species leave the column totally separated the feed pulse is too small for total utilization. When there is too much overlap of the various species at the column exit and purity is low for reasonable recovery levels then the column is overloaded. A suitable compromise between these extremes is looked for.

Feed concentration or % dry solids is especially important in the fractionation work since diluted feeds bring unwanted liquid volume into the system and limit the separation efficiency as well as requiring a larger system. Too high a feed concentration can also overload the resin and result in losses of the species that has the higher affinity for the resin. This can be corrected through adjustment of the resin flowrate.

Elution velocity can have a significant effect on the design when the HETP is not linearly dependent on the fluid velocity. Capital costs are more sensitive to column diameter than they are to column height (especially in the case of the ISEP). Consequently, there is a strong incentive to operate at higher fluid-flow velocities. As previously mentioned, there is a basis for the relationship,

$$\text{HETP} \propto v_{\text{fluid}}^{0.5}$$

which provides us with a strong incentive to maximize the velocity as long as pressure drop, dispersion effects and other factors do not begin to offset the sorbent productivity gains realized from the above relationship.

Eluent temperature is always a variable to investigate during preliminary productivity testing and can have much influence on the separation factor. Higher temperatures usually bring better kinetic performance but not always an improved interspecies separation.

CHROMATOGRAPHY EXPERIMENTAL PROCEDURES

The laboratory pulse test is the standard procedure for evaluation of chromatographic separations. Information from these tests is sufficient to estimate conceptual performance and sizing for commercial operations.

In setting up a test for chromatography the attention to experimental detail and the control of the test environment is more critical than for the previously described adsorption testwork. This is simply because adsorptive separations are more robust, operate at much higher values of α, the separation factor, and often are accompanied by external factors that influence more the conversion efficiency, more so than does the condition of the sorbent bed. The following areas must usually be addressed prior to the experimental runs.

Resin pre-conditioning is done slowly and with excess eluent so as to eliminate any residual chemical impurities that might still remain on the manufactured product. All pre-conditioning is performed with solutions prepared from de-aerated water and resin beads are not then left exposed to air where they might get air into the porous structure again.

Resin loading is a time-consuming process and must be done slowly by loading small batches of a well-mixed resin sample. The small resin batches are slurried into a water-filled column and allowed to settle on top of previous batches. The introduction of each batch is done in a manner such that air entrainment is not allowed. The column is not upflowed during this or any later preparation effort since we do not want any elutriation effect to bias the perfomance.

De-aerated fluids must be employed in the conduct of chromatography especially when higher temperatures are used. Degassed fluids at room temperature will often release dissolved gas at higher temperature when in contact with a nucleating surface such as a resin bead and therefore the de-aeration must be done at a suitably high temperature. Gas release causes accumulation in the resin pores to occur and it can be difficult to remove air from hydrophobic resin surfaces with aqueous liquids. As gas accumulates pore volume decreases, sites become unavailable to the liquid phase and the distribution coefficients, K_ds, will change, thereby affecting the separation efficiency.

Column packing is important. Recommendations for this are to run eluent through the bed for several hours while monitoring the pressure drop, δP. Initially the δP will increase and then stabilize. The fluid flow rate in this column packing operation should be representative of the subsequent flow rate.

Preliminary separation runs are recommended to stabilize the column to the chemical cycle it will experience during testing and production. Three or four runs are normally sufficient to provide repeatable results. Special attention is required when the sorbent undergoes volume change during the chromatography cycle.

Method of feed addition can be critical to losses resulting from unnecessary dispersion as the feed enters the bed as eluent follows the feed pulse. The larger the pulse the less sensitive the separation efficiency depends on the feed and eluent addition technique. It is important for the same reason to maintain a 'good' level profile at the surface of the bed.

The valuable information gained from conducting experiments as described above are:

Column productivity data which will be the major factor in choosing the conditions for operating in the continuous chromatography multiple component separation mode.

Effluent profiles are plotted to obtain curves as illustrated in Fig. 10.33 from which, as shown on the figure, the HETP values can be estimated for the conditions of the test. Capacity factors, k', are also calculated from these pulse-test effluent profiles.

$$\text{HETP} = L/P, \qquad L = \text{Column Length}$$
$$P \equiv \text{Number of stages} = 2V_{max}[V_{max} - V_0]/W^2$$
$$k' = [V_{max} - V_0]/V_{max}$$

The experimentation designed to generate design data for fractionation manipulates much the same group of variables. The fluid-velocity data for HETP estimation is then directly from the pulse tests. The single most different test is the isotherm data generation which is performed in an identical manner to that described for adsorption testing in the Adsorption Process Design Procedures section.

Fractionation equipment size decreases as phase capacity or solute solubility increases. Therefore there is a large incentive to operating at high solute concentration provided there are no severe equilibrium and kinetic disadvantages. The effect of higher temperatures usually alleviates the kinetic problem through increased diffusion rates.

Reflux is useful in maintaining high effluent concentrations and high sorbent use. Reflux does not normally decrease the size of the cascade to any great degree. Reflux should not materially increase the fluid flows either, provided the solute is refluxed at increasing fluid phase concentration. The same arguments apply to reflux re-entering the fractionation zone as reloaded resin.

Fractionation design must address the elution of the adsorbed species. This operation is akin to an elution of the type described in the Adsorption Process Design Procedures section, and is designed in a similar fashion. It is possible here to design a special eluent system followed by a sorbent re-conditioning step. This could be a unique advantage of the ISEP over other fractionation systems.

CHROMATOGRAPHIC PROCESSES

The first work executed by AST in chromatography was the separation of light rare-earth elements. It was accomplished using the multiple-component separation technique referred to as continuous chromatography above. The tests were successful but did not lead to any commercial exploitation of the technology. The arrangement for multiple fractionation work is shown in Fig. 10.31. The initial work has not included the mid-fraction recycle.

Isomer separation

A recent project with a pharmaceutical application is likely to lead to commercial practice. The feed is a mixture of sugars. Two very similar saccharide molecules, A and B, dominate the mixture and are required to be separated such that the one with most affinity for the resin, B, is desired at high purity and recovery. Initially laboratory pulse tests determined the system capacity factors (k' values) at different temperatures and mixture ratios and also attempted to find the optimum column operating conditions. The k' factors as a function of temperature are listed in Table 10.7. Comparison of separation factor values, $\alpha = k'_B/k'_A$ shows a distinct advantage in operating a separation at 70°C where it is a maximum.

Prior to running the continuous ISEP we proceeded to run parallel bench tests to establish a baseline performance for 1 m, 3 × 1 m and 3 m deep batch columns. These two lengths were chosen as potential actual depths for the commercial facility. The baseline performance would provide criteria on which to judge the ISEP continuous run data. The data for the 1 m column was inconsistent and was scheduled to be rerun. Figs 10.37 and 10.38 are graphical representations of the 3 × 1 m and 3 m column tests' effluent histories — summarized in Table 10.7.

A test was also run on a group of columns arranged in series — 3 × 1 m. The data from this test indicated that such an arrangement is inferior to the single 3 m column.

Continuous adsorption and chromatography

Table 10.7

	3 × 1 m column	3 m column	ISEP
Feed pulse (BV)	0.8	0.8	0.8
Elution velocity (BVs/hr)			
Elution time (minutes)			
Total cycle time (minutes)			180
Productivity (g/min/litre)			
Recovery (desired component)			74
Purity (desired component)			98

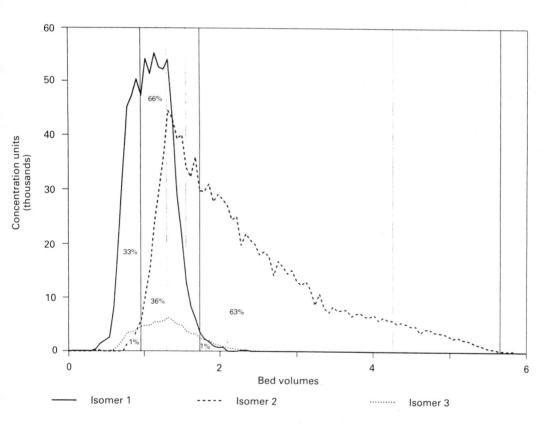

Fig. 10.37 — Sugar isomers separation (3 × 1 metre column elution).

The 3 × 1 m experiment was supposed to emulate a single 3 m column. If results had been satisfactory, such a physical arrangement could have been set up on the ISEP.

The ISEP contactor was then prepared to operate under identical conditions to those employed in the 1 m column test. The ISEP cycle time was 3 hours which is the

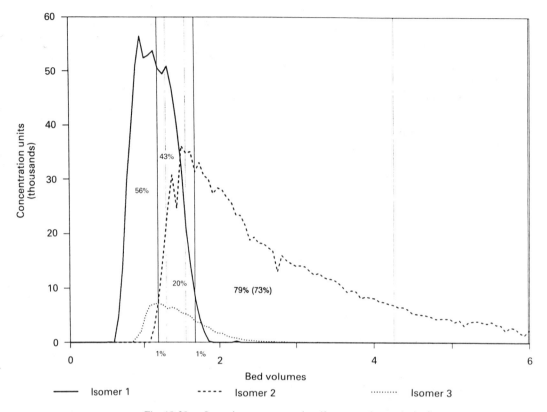

Fig. 10.38 — Sugar isomers separation (3 metre column elution).

same as the batch column cycle time. The ISEP configuration is shown in Fig. 10.38. The whole unit was insulated to maintain temperature and all feed and eluent was pre-heated. Temperature loss across the 1 m columns was approximately 10°C. This is considered high but will be much less in a commercial column of larger diameter.

The ISEP results were duplicated in a second test and the results are shown in Fig. 10.39. The ISEP performance shows that the effluent resolution capability over 20 ports is sufficient to get equivalent performance, as was seen in the static column. The mixing of fractions that occurs when two columns discharge into the same outlet port (see Fig. 10.40) does not seriously influence the separation quality. This mixing in the discharge will become more critical as the number of different fractions desired increases. In such cases a valve with more ports is recommended. The actual ISEP performance with 1 m columns equalled the static 3 m column. This is a surprising and welcome result since this means that a 1 m bed can do the same job with one third the resin. It will execute three smaller pulsed cycles in the same time a 3 m column will process a larger pulse.

High recovery of the desired component at a purity of 98% requires that a recycle stream of mixed product be combined with the feed, as shown in Fig. 10.41. This phase has not yet been executed. The net result will be to increase the recovery of B

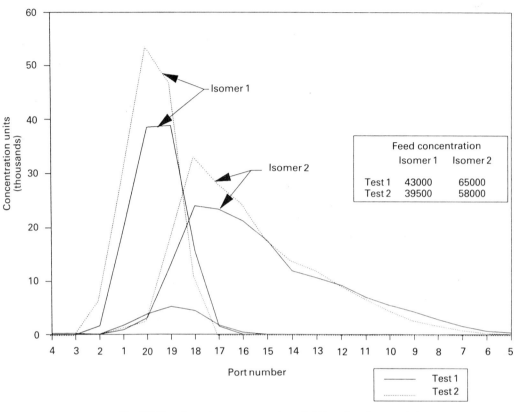

Fig. 10.39 — Sugar isomers separatioon (L100 a pilot plant results).

up to a maximum, increase the mass feed flow to the ISEP separator to some percentage above the net new feed flow and to increase the size of the ISEP separator chambers. The actual increase expected will depend on how different the combined feed mixture is relative to the raw feed composition.

Equilibrium data tests
The next phase of the isomer separation will return to the laboratory and develop the isotherms at the various conditions that exist in the fractionation operation. The assumption of linear, constant K_d values is usually incorrect. Deviations from linear equilibrium lines can have dramatic effects on the operating line (phase flow ratios) constraints and on the number of stages needed to achieve the separation. The isotherm examples in the Adsorption Process Design Procedures section demonstrated this aspect of phase-equilibria relationships.

Fractionation runs
Fractionation runs are underway to demonstrate the efficiency of operating in this mode. Results from this phase will be available at a later date. The criteria for success will be the comparison with the previous circuit arrangement.

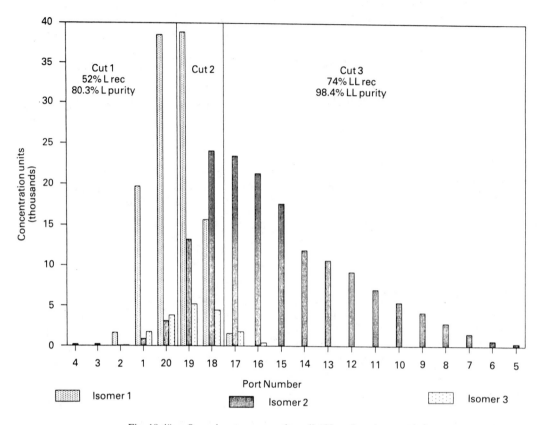

Fig. 10.40 — Sugar isomers separatioon (L100 a pilot plant results).

Ch. 10] Continuous adsorption and chromatography 223

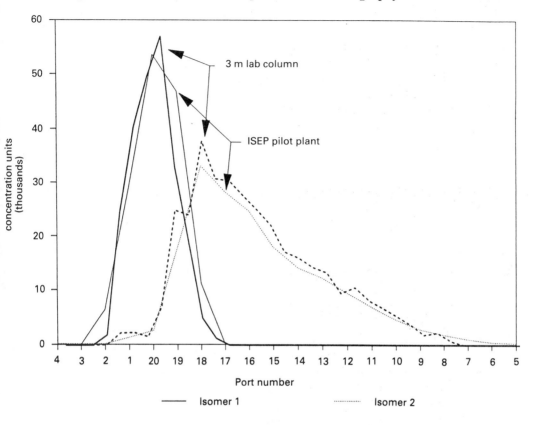

Fig. 10.41 — Sugar isomers separation (comparison of lab. and pilot plant results).

REFERENCES

[1] Olof Samuelson, *Ion Exchange Separations in Analytical Chemistry*, John Wiley, New York, 1963, p. 105.
[2] D. W. Simpson and R. M. Wheaton, *Chemical Engineering Progress*, **50**, 47.
[3] R. J. W. Wooley, Modeling the continuous ion exchange (ISEP) process using ASPEN/SP, *1990 AIChE Annual Meeting, Nov. 11–16, 1990, Chicago.*
[4] Barker, *et al.*, Theoretical aspects of a preparative continuous chromatograph, *Chromatographia*, **10** (1977).
[5] N. Benedict, T. H. Pigford and H. W. Levi, *Nuclear Chemical Engineering*, 2nd edn, McGraw-Hill, New York, 1981, pp. 183, 184.
[6] T. C. Lo, M. H. I. Baird and C. Hanson, *Handbook of Solvent Extraction*, John Wiley, New York, 1983, pp. 178, 179.

11

Applications of liquid chromatography to large-scale purification of bacterial proteins

Christopher R. Goward
Centre for Applied Microbiology and Research, Porton Down, UK

INTRODUCTION

Laboratory-scale purifications are designed to produce a highly purified product without emphasis on cost or maximum recovery of product. The quantity of sample is small and the aim is to achieve the maximum separation and gain highly purified protein; flow rates are relatively low because resolution is of greater importance than throughput. On the large scale the cost of production compared with the value of the product is of prime importance. The maximum amount of protein is prepared in the minimum time; flow rates are high because throughput is relatively more important than resolution. Large-scale conditions may be needed when the protein is present in low concentration or when a large amount is needed. The degree of purity required depends partly on whether the protein is to be used as diagnostic reagent or as a therapeutic agent and partly on the undesirable activity of any contaminating protein. A therepeutic protein has to be pure to have minimal side-effects when administered to a patient, but a diagnostic protein need not necessarily be pure.

The choice of the source of protein is determined by the nature of the product but it may be animal, plant or microbial. Bacteria, in general, have a greater range of activities than animal cells. Animal sources are cheap but their enzyme content is relatively static with some variation bacterial cells have greater potential, particularly in that protein production can be increased by cultural and genetic techniques; they can be cultured on almost any scale and a specific gene can be cloned and placed in front of a strong promoter to gain high expression.

Genetic engineering and cloning means bacteria can be forced to produce typically 5–20% (and sometimes up to 50%) total soluble cell protein as a single protein such that processes which may have required 50 kg cell paste may only now require 5–10 kg cell paste containing the recombinant gene. Unfortunately not all genes for proteins can be readily cloned.

Most industrial enzymes are extracellular and need little purification. Enzymes used for other purposes are often intracellular and are produced on the large scale in up to several thousand litres of culture. Intraculluler proteins have to be released from the cell and extracellular proteins are recovered from culture supernatant after a concentration step. The bacterial cells may be disrupted by use of chemicals, such as detergents, alkali or enzymes, or by physical methods such as liquid shear, solid shear, agitation with abrasives, freeze-thaw or osmotic shock. The solid and liquid portions are separated by centrifugation or filtration. It may be necessary to remove nucleic acids by treatment with nuclease or by precipitation. Partial purification of the target protein may be achieved by precipitation with salts, heat or by a change of pH, by aqueous two-phase separation, or by chromatography. These purification methods may be combined in a process but the purpose of this chapter is to concentrate on aspects of chromatography relating to large-scale purification of bacterial proteins. The amount of a protein required for therapeutic or diagnostic use may be small so the production of low-volume but high-value products by chromatography becomes viable [1].

MATRICES

In the past it was common to produce matrices in the laboratory but now most matrices, other than some affinity matrices, are commercially available. The correct choice of matrix is a most important consideration. It is desirable that the matrix has certain properties. It should be hydrophilic, macroporous (microporous for gel filtration) and the particles should be mechanically rigid to allow flow rates — the particles should also be small, uniform and spherical to confer good flow properties. The matrix should be easily derivatized with various functional groups, be resistant to chemical and biological degradation, inert for minimum non-specifc adsorption and should be reusable. No matrix has all these properties so a range of matrices are available, most in derivatized form, and have recently been reviewed by Atkinson *et al.* [2]. Cellulose, dextran and agarose have insufficient rigidity at porosities suitable for protein chromatography. Newer matrices have a more rigid structure such as the cross-linked derivatives bascd on the above polymers, for example Sepharose Fast Flow, or synthetic matrices such as Triacryl [3]. The hydrodynamic properties of these newer matrices make them suitable for large-scale work where high flow rates are desirable. These more rigid matrices may also be regenerated without their removal from the column since their volumes remain constant. Inorganic materials such as porous silica have not been widely used. While rigid, they tend to denature or bind proteins irreversibly even after derivatization, and this has limited their use [4]. To some extent the problem of compressibility can be overcome using a composite of kieselguhr coated internally and externally with agarose. This composite, macrosorb (Sterling Organics, Newcastle), combines the hydrodynamic advantages of a rigid, inorganic material with biological advantages of agarose, as described for the purification of the therapeutic enzyme L-asparaginase from *Erwinia chrysanthemi* [5]. It can be derivatized for use in ion-exchange, hydrophobic interaction chromatography and affinity chromatography and it can be used in columns but is probably more applicable for use in fluidized beds.

PARTICLE SIZE

Although at low flow rates there is little overall difference between small and large particles, the best resolution and thoughput is obtained using small particles (<10 μm diameter) and high flow rates but this leads to a high back pressure such that the flow rate may have to be reduced. Small particles are more expensive as is the high-pressure equipment required; a compromise is to use larger particles (30 μm–50 μm diameter) at the highest flow rate compatible with the gel, the resolution required and the equipment used. The new generation of matrices are based on the early polymers but are cross-linked or made of composites. They have high rigidity and are less compressible and more uniform with smaller particle size such that these matrices are more suitable for large-scale. Monobeads, Superose and the TSK-PW range (ToseHaas, Philadelphia, USA) all have 10 μm particles capable of operation at 3–10 MPa. More important for low-pressure liquid chromatography is Sephacryl HR, Sepharose, FF, Superose and the Fractogel range (TosoHaas) which are in the 40 μm range. These gels give higher resolution than conventional gel beads of 90 μm or more.

THE PRINCIPLES OF CHROMATOGRAPHY

The principles of chromatography have been described in the literature, for example by Scopes [6] and scale-up to large scale has been reviewed by Janson and Hedman [7] and Scawen and Melling [8,9]. The major categories of chromatography are briefly outlined below.

Ion-exchange chromatography

Ion-exchange is the most generally useful step in large-scale chromatography and is based on the net charge of the target molecule. The matrix is usually relatively inexpensive compared to, for example, an affinity matrix. An ion-exchange matrix is substituted with charged groups and associated with mobile counter-ions which can be reversibly exchanged with ions of the some charge. The ion exchanger may be positive and have negative counter-ions (anion exchanger) or it may be negative and have positive counter-ions (cation exchanger) and examples of common substituents are shown in Table 11.1.

The capacity of ion exchangers is high, sample application is not limited by volume and protein can be adsorbed form a large volume. Resolution is high and the flow rates can be medium to very high and the method can be applied to any scale. It is often used at an early stage in a process to reduce the volume of the sample. Conditions for binding to the elution from an ion-exchange matrix is dependent upon ionic strength, pH and temperature. Elution from an anion exchanger is with a decreasing pH gradient or an increasing gradient of ionic strength. Elution from a cation exchanger is with, again, an increasing pH gradient or an increasing gradient of ionic strength.

An example of the power of ion-exchange chromatography can be seen with the purification of β-lactamase from *Enterobacter cloacae* [10]. The process involves separation of the cell extract by cation-exchange chromatography on CM-Sepharose

Table 11.1 — Common ion-exchange substituents

Cation-exchange substituents	
Carboxymethyl (CM)	$-O-CH_2-COO^-$
Sulphoethyl (SE)	$-O-CH_2-CH_2-SO_3^-$
Sulphopropyl (SP)	$-O-CH_2-Ch_2-CH_2-SO_3^-$
Sulphonate (S)	$-O-CH_2-SO_3^-$
Phosphate (P)	$-O-PO_3^-$

Anion-exchange substitutes	
Diethylaminoethyl (DEAE)	$-O-CH_2-CH_2-N^+(C_2H_5)_2$
Quatenary aminoethyl (QAE)	$-O-CH_2-CH_2N^+-(C^2H_5)_3$
Quaternary amine (Q)	$-CH_2-N^+-(CH_3)_3$

FF. The protein is applied to the column and unbound protein washed in 10 mM potassium phosphate buffer, pH 6.0. Pure β-lactamase is recovered in this single step after elution with a linear gradient of 10–100 mM potassium phosphate buffer, pH 6.0.

Affinity chromatrography
Affinity chromatography is a method in which the target molecule is specifically and reversibly adsorbed to a ligand immobilized on a matrix. Ligands used for affinity chromatography may be substrate, substrate analogue, co-factor, antibody, inhibitor, receptor, dye or antigen. Unbound protein is washed from the column and the target protein eluted using the mildest conditions available. Elution may be selective by addition of a biospecific agent (substrate, co-factor, inhibitor, free ligand) or it may be non-selective by altering the ionic strength, the pH, polarity (ethylene glycol, surfactant), or addition of a denaturing agent or a chaotropic agent (KSCN). Biospecific elution is normally expected to give the best purification but is often very expensive for large-scale work. It is important the biospecific interaction is reversible ($K_d=10^4$ to $10^8 M^1$). Immunoaffinity is becoming more widely used as monoclonal antibodies are becoming more available. Affinity chromatography is useful in that very high resolution can be obtained, the process time may be rapid and the overall number of steps in a process may be reduced. Disadvantages are that the ligand may be expensive, unstable and may leach from the matrix, which may cause problems if the ligand is expensive, and the product may be contaminated. The product has to have a high value and the derivatized matrix may be difficult to synthesize. The capacity of affinity media can be low or high, but their operation is not limited by the volume applied. Resolution may be very high and the speed of operation may also be high. This system is best used at a later stage in a process to protect the matrix. The method is very powerful but has not found widespread use on the large scale owing to the disadvantages described above.

The ligand is immobilized to a matrix, usually agarose. Chemicals used to immobilize the ligand include cycanogen bromide, divynyl sulphone, tresyl chloride and bis-epoxides. All have advantages and disadvantages but cyanogen bromide is the most often used [11]. Coupling the often-labile ligand to a matrix is difficult and the coupled ligand is often unstable and open to biological attack. Reactive dyes offer an alternative ligand. These dyes are mainly produced by ICI (Procion) and CIBA (Cibacron). They consist of an anthroquinone, diazo or phthalocyanine chromophore linked to a reactive triazinyl group. This reactive group is coupled to the matrix at mildly alkaline pH and room temperature, although the dye has recently been immobilized via the chromophore with very encouraging results [12]. The immobilized dye is stable, although prone to leakage from the matrix, and binds a wide range of proteins under specific conditions [13].

An example of the simplicity and power of affinity chromatrography is illustrated by the purification of recombinant protein G expressed by *E. coli*. Cell extract is applied to IgG-Sepharose equilibrated with 50 mM Hepes-NaOH, pH 8.0 containing 250 mM NaCl. Unbound protein is washed from the column with the same buffer and pure protein G eluted with 100 mM glycine-HCl, pH 2.0. Any proteolytic fragments of the protein G molecule can be removed, if necessary, by anion-exchange chromatography on Q-Sepharose FF [14].

Hydrophobic-interaction chromatography
Hydrophobic-interaction chromatrography separates proteins based on their differing relative strengths of hydrophobic interaction with a matrix substituted with hydrophobic groups. The capacity of these matrices is high and adsorption of protein is not limited by the volume of sample. Proteins usually bind at high ionic strength so this step is best positioned after a salt precipitation or ion-exchange chromatography. Hydrophobic interations are increased and binding promoted by increasing the concentration of salting-out ions, lowering the pH, raising the temperature or increasing the hydrophobicity of the ligand. Hydrophobic interactions are decreased and elution affected by lowering ionic strength, increasing polarity , adding a detergent, raising the pH, reducing the temperature, increasing the concentration of chaotropic ions, introducing specific deformers of the protein or simply by adding water.

The resolving power of hydrophobic-interaction chromatography may be illustrated by the purification of glucokinase and glkycerokinase from *Bacillus stearothermophilus*. The enzymes could be separated by a procedure which included three dye affinity chromatography steps following partial separation by anion-exchange chromatography [15,16]. A drawback with the dye affinity chromatography method was contaminated of the product with leached dye. The proteins proved difficult to separate by methods which did not use dye affinity chromatography but could be obtained free of each other by one step on phenyl-Sepharose [17]. The cell extract was initially separated by anion-exchange chromatography on DEAE-Sepharose FF, the extract being applied in 100 mM potassium phosphate buffer, pH 8.0. The column was washed to remove unbound protein and eluted with a linear gradient of 100–400 m potassium phosphate buffer, pH 8.0. Glucokinase and glycerokinase co-elute and were then applied to the hydrophobic interaction chromatography matrix phenyl-Sepharose CL-4B in 400 mM potassium phosphate buffer. The column was washed with the same buffer and glucokinase eluted with a linear gradient of 400–10 mM potassium phosphate buffer, pH 8.0. Glycerokinase remained bound but could be eluted with water. The enzymes were finally polished separately by gel-filtration chromatography on Ultrogel AcA 34 in 650 mM potassium phosphate buffer.

Gel-filtration chromatography
In gel filtration molecules are separated according to differences in their molecular size and shape. Large molecules are unable to enter the pores of the matrix and so pass straight through, whereas smaller molecules enter the pores and are eluted in order of decreasing size. There should be no interaction between the matrix and the solution so the buffer medium should be inert. The matrix should be rigid, as it determines flow rate, and very porous. The method is rarely used for large-scale fractionation; its main use is to separate proteins from salt or other substances. The capacity of gel-filtration media is low and limited by the volume of sample, <5% for fractionation and 30–40% for desalting. Flow rates are low for fractionation and medium for buffer exchange. The resolution is low-to-medium for fractionation and its best carried out at the end of the process when the sample volume is relatively low. The method may be used for buffer exchange at any stage although its application may be limited by the volume of the sample.

Chromatofocusing and immunoaffinity chromatography may be used on the large-scale but they are expensive so are usually restricted to laboratory-scale separations.

PROCESS DESIGN

The techniques mentioned above the applicable to use where 10–50 kg cell paste is processed, or up to several thousand litres of culture supernatant, to produce less than a kilogram of product. Large-scale chromatography steps should include high resolution to minimize the number of steps in the process, high capacity to minimize the quantity of matrix required, speed to minimize the cycle time, a concentration step to handle large sample volumes and an easily regenerated system with minimum loss of process time between cycles. Process economy is of prime importance since there is no point in doing unnecessary steps. The order of steps needs careful consideration to make a process efficient. The first step has to be capable of handling, and preferably reduce, a large volume. It is important for process economy to minimize the change in conditions, use concentrating steps after diluting steps and use gel filtration at the end if it is necessary. The first concentration step is often ion exchange. Batch procedures have often been used in the past but this is changing as new matrices with high flow rates facilitate use in a column procedure. The use of a column gives better resolution and greater purification than can be achieved with batch elution. Hydrophobic interaction chromatogrphy may also be used early in a process as it too has the advantage of concentrating the solution and often removes unwanted proteolytic enzymes. High ionic strength is usually used for adsorption and low ionic strength for elution; this is in contrast to ion-exchange chromatography so the two methods can be interchanged as appropriate. Affinity chromatography media may be susceptible to fouling if used at an early stage in a process and may be difficult to regenerate which is of particular note owing to their high cost, but the method may be useful at an early stage to extracellular proteins. Gel-filtration chromatography is usually used only as a polishing step and is of limited use on the large scale, although it is sometimes used for buffer exchange.

SCALE-UP

The process should be fully optimized in the laboratory and the sample volume may then be increased by 10 or 20 times. The column surface area is increased proportionally with the bed height constant; the column diameter has little effect on resolution so the amount of matrix used can be increased by increasing the diameter of the column while keeping the column length constant. The gel type, ionic strength, pH, linear flow rate and temperature should remain the same although the flow rate may be increased if greater throughput is required. Large-scale cell extracts often contain higher protein concentrations than small scale owing to differences in cell-breakage methods, so care has to be taken not to overload the column. The scale-up process is repeated until the desired scale is achieved.

CHOICE OF COLUMN

Column materials should be compatible with solvents and solutions used. Short, board columns are best for ion exchange, hydrophobic interaction and affinity chromatography although there are occasions where a longer column can be exploited to resolve proteins further. Longer columns are required for gel filtration. The maximum column diameter commonly available is 120 cm; with larger columns it is difficult to ensure even loading over the entire surface but two or more columns may be run in parallel to resolve this. Soft gels may also be used in such a 'stack' system which reduces the compression of gel. Bed heights are commonly 15–25 cm for ion exchange, which is the most commonly used matrix in large-scale purifications. Dead volumes should be similar to those used in the laboratory scale. Jansson and Hedman [7] have detailed desirable features of large-scale chromatrography columns. Dimensions of the column become more critical with process chromatography compared with laboratory chromatography because of problems with back pressure which limits reduction in process time.

MAINTENANCE OF COLUMNS

Effective cleaning is required to ensure longevity of the matrix and its continued correct function in order to reduce the risk of contaminating the product with endotoxins or other unwanted bacterial products. Fouling of the column may be from particulate matter, non-specifically adsorbed substances or from microbial contamination. Particulate matter including micro-organisms may be removed with a filter upstream of the column. Non-specifically adsorbed substances may be removed from the column by washing with 2M NaCl, up to 6M urea, non-ionic detergents (1% Triton X-100) or up to 2M NaOH, either singly or in combination. To prevent growth or to remove micro-organisms or pyrogens washing with 0.5–2M NaOH is preferred. Affinity-chromatography matrices may present a greater problem if the ligand is labile but a high concentration of salt may be passed through the column or moderately low and high pH washes (pH 4.5 and 8.5). Autoclaving sterilizes a matrix but has little cleaning effect, endotoxins may not be destroyed and the procedure is not possible to attain in a packed column. The matrix may be stored in sodium azide or merthiolate but these bacteriostatic agents may allow a low level of bacterial contamination and cannot be used with all affinity media, they have no cleaning effect and there is the risk of contamination of the product and so are not suitable for therapeutics. NaOH is the best choice, where applicable, because is sterilizes the entire system, has good cleaning effect, cannot contaminate the product and it destroys pyrogens, but it does require careful handling on the large scale. Ethanol is useful for storing the matrix and can be used for reducing bacterial contamination.

THE FUTURE: PROTEIN ENGINEERING APPLIED TO PROTEIN PURIFICATION

Through the 1980s molecular genetics has produced recombinant systems so that bacteria can now produce heterologous proteins in high concentration with yields. Inclusion bodies are often formed and after cell lysis the insoluble aggregates may be isolated as substantially pure proteins that only require solubilization with urea and then careful renaturation since incorrect conformation may lead to a decrease in biological activity. The gene can also be manipulated to produce proteins with more desirable properties including fusion of the genes producing peptides to the target protein to aid protein purification. DNA sequences which codes for a polypeptide containing desirable properties to aid purification of the target molecule may be introduced upstream or downstream of the portion coding for the product. The purification tags have been used to exploit affinity, ion exchange, hydrophobic interaction, metal chelate and covalent separations and may be applicable to large-scale purifications. Examples of affinity tags are summarized in Table 11.2.

The target molecule with the fused affinity tag can be separated by adsorption to an appropriate ligand as shown in Table 11.1. Non-bound protein is washed through the column and the target protein eluted by changing solutions of pH, ionic strength or addition of a competing substance. The polypeptide tail may be released after chromatography by chemical or enzymic means. The released tail, if intact, may be captured by passing the mixture through the same chromatographic matrix or separated by size or charge, but removal of the affinity tail, if necessary, is often the most difficult problem [18]. Examples of the cleavage methods which may be used to remove affinity tags are shown in Table 11.3. The big advantage of the method is that rather than screening matrices and optimizing the process the protein can tailored to suit a purification method.

Table 11.2 — Examples of affinity tags

Tag	Ligand
Antigenic peptide	Monoclonal antibody
Arginine	Anion exchange
Avidin/streptavidin	Biotin
β-Galactosidase	β-D-Thiogalactoside
Chloramphenicol acetyl transferase	Chloramphenicol
Cysteine	Thiol
Glutamate/aspartate	Cation exchange
Histidine	Cu^{2+}, Ni^{2+}, Zn^{2+}
Phenylalanine	Phenyl or Octyl
Protein A	IgG
Protein G	OgG or albumin

Table 11.3 — Specific cleavage methods used to remove tags. The position of the cleavage site is marked (▼)

Cleavage site	Reagent
–Asn▼Gly–	Hydroxylamine
–Asp▼Pro–	Weak acid
C-terminal ▼Arg or ▼Lys	Carboxypeptidase B
C-terminal residues (not Arg or Lys)	Carboxypeptidase A
–Cys▼	2-Nitro-5-thiocyanatobenzoate
–Gly–Val–Arg–Gly–Pro–Arg▼	Thrombin
–Ile–Glu–Gly–Arg▼	Factor X_a
▼Ile–Val–Gly–Gly–Thr–Val–	Factor XII_a
–Met▼	Cyanogen bromide
–Pro–Xxx▼Gly–Pro–	Collagenase
–Trp▼	N-Bromosuccinamide

Some examples of the use of affinity tags demonstrate the versatility of the techniqiue. Polyarginine tails to the C-terminus of recombinant urogastrone makes the protein basic and bind strongly to cation-exchange matrices at pH 5.5 where 90% of *E. coli* proteins pass straight through the column [19,20]. The polyarginine tail can be digested with carboxypeptidase B and the protein repurified on the same ion-exchange matrix where it moves the elution position relative to other proteins. Eleven Phe residues fused to the N-terminus of β-galactosidase enabled the protein to be separated on phenylo-Superose [21]. Four Cys residues fused to galactokinase facilitated separation on thiopropyl-Sepharose [21]. Mouse dihydrofolate reductase tagged with a polyhistidine peptide could be purified using metal chelate chromatography followed by removal of the affinity peptide with carboxypeptidase A [22]. A short hydrophilic antigenic peptide of eight residues engineered on to the N-terminus of recombinant lymphokines allows them to be purified using an immobilized monoclonal antibody specific for the first four amino acids of the peptide [23]. A fusion of two IgG-binding domains of protein A to recombinant human insulin-like growth factor I (IGF-I) [24], expressed in *E. coli* and *Staphyloccus aureus*, can be separated on IgG-Sepharose. The fusion protein was secreted from *E. coli* into 1000 litres of culture medium and bound to IgG-Sepharose at neutral pH then eluted at low pH on a large scale [25]. Treatment with hydroxylamine released the protein by cleaving an Asn–Gly bond. Disadvantages of such protein A fusions are then extreme conditions of pH or high concentrations of chaotropes required to elute them may also denature the desired protein. There may also be difficulty in subsequent removal of the protein A moiety without damaging the desired protein. The second point is common to other fusions. An acid-labile bond may be introduced at the junction between the species of protein otr a protease-sensitive site introduced in the same region. For example, introduce Asn–Gly bonds which are rare and can be broken with hydroxylamine [26]. Any Asn–Gly bonds in the required protein product may be removed by protein engineering without hopefully affecting the properties of the protein. A dual-affinity system has been reported where tails are fused to the N-terminus and the C-terminus which aids stability of the target molecule and allows two purification methods to be exploited. An example is the fusion of the albumin binding domains of protein G, and IgG binding domains of protein A around as recombinant human insulin-like growth factor II [27]. It is likely that these types of methods will be incorporated in the separation of bacterially produced proteins in the future.

REFERENCES

[1] P. M. Hammond and M. D. Scawen, 'High-resolution fractionation of proteins in downstream processing', *J. Biotechnol.*, **11**, 119–134 (1989).

[2] T. Atkinson, M. D. Scawen and P. M. Hammond, 'Large-scale industrial techniques of enzyme recovery', in *Biotechnology*, H.-J. Rehm and G. Reed (Eds), Vol. 7A, VCH Verlagsgesellschaft, Weinheim, pp. 279–323 (1987).

[3] T. Miron and M. Wilchek, 'Activation of Trisacryl gels with chloroformates and their use for affinity chromatography and protein immobilisation', *Appl. Biochem. Biotechnol.*, **11**, 445–456 (1985).

[4] S. Berkowitz, 'Silica-based chromatography products', *Bio/Technology*, **5**, 61–62 (1987).

[5] C. R. Goward, G. B. Stevens, I. J. Collins, I. R. Wilkinson and M. D. Scawen, 'Use of Macrosorb kieselguhr composite and CM-Sepharose Fast Flow for the large-scale purification of L-asparaginase from *Erwinia chrysanthemi*', *Enzyme Micronb. Technol.*, **11**, 810–814 (1989).

[6] R. K. Scopes, *Protein Purification: Principles and Practice*, 2nd edn, Springer-Verlag, New York (1987).

[7] J. C. Janson and G. Hedman, 'Large-scale chromatography of proteins', *Adv. Biochem. Eng.*, **25**, 43–99 (1982).

[8] M. D. Scawen and J. Melling, 'Large-scale extraction and purification of enzymes and other proteins'. in *Handbook of Enzyme Biotechnology*, 2nd edn. A. Wiseman (Ed.), Ellis Horwood, Chichester, pp. 15–53 (1985).

[9] M. D. Scawen and J. Melling 'Practical aspects of large scale protein purification', in *Handbook of Enzyme Biotechnology*, A. Wiseman (Ed.), 2nd edn, Ellis Horwood, Chichester, pp. 247–273 (1985b).

[10] C. R. Goward, G. B. Stevens, P. M. Hammond and M. D. Scawen, 'Large scale purification of the chromosomal β-lactamase from *Enterobacter cloacae* P99', *J. Chromatogr.*, **457**, 317–324 (1988).

[11] R. Axen, J. Porath and S. Ernback, 'Chemical coupling of peptides and protein to polysaccharides by means of cyanogen halides', *Nature (Lond.)*, **214**, 1302–1304 (1967).

[12] S. J. Burton, C. V. Stead and C. R. Lowe, 'Design and applications of biomimetic anthraquinone dyes. III. Antraquinone-immobilized C.I. Reactive Blue 2 analogues and their interaction with horse liver alcohol dehydrogenase and other adrenine nucleotide-binding proteins', *J. Chromatogr.*, **508**, 109–125 (1990).

[13] M. D. Scawen and T. Atkinson, 'Large scale dye ligand chromatography', in *Reactive Dyes in Protein and Enzyme Technology*, Y. Clonis, T. Atkinson, C. J. Bruton and C. R. Lowe (Eds), Macmillan, Basingstoke, pp. 51–85 (1987).

[14] C. R. Goward, J. P. Murphy, T. Atkinson and D. A. Barstow, 'Expression and purification of a truncated recombinant streptococcal protein G', *Biochem. J*, **267**, 171–177 (1990).

[15] M. D. Scawen, P. M. Hammond, M. J. Comer and T. Atkinson, 'The application of triazine dyes to the large-scale purification of glycerokinase from *Bacillus sterothermophilus*, *Anal. Biochem.*, **132**, 413–412 (1983).

[16] C. R. Goward, R. Hartwell, T. Atkinson and M. D. Scawen, 'The purification and characterisation of glucokinase from the thermophile *Bacillus stearothermophilus*', *Biochem. J.*, **237**, 415–420 (1986).

[17] C. R. Goward, T. Atkinson and M. D. Scawen, 'Rapid purification of glucokinase and glycerokinase from *Bacillus stearothermophilus* by hydrophobic interaction chromatography', *J. Chromatogr.*, **369**, 235–239 (1986).

[18] H. M. Sassenfeld, 'Engineering proteins for purification', *TIBTECH.*, **8**, 88–93 (1990).

[19] H. M. Sassenfeld and S. J. Brewer 'A polypeptide fusion designed for the purification of recombinant proteins', *Bio/Technology*, **2**, 76–81 (1984).

[20] S. J. Brewer and H. M. Sassenfeld, 'The purification of recombinant proteins using C-terminal polyarginine fusions'. *Trends Biotechnol.*, **3**, 119–122 (1985).

[21] M. Persson, M. G. son Bergstrand, L. Bulow and K. Mosbach, 'Enzyme purification by genetically attached polycysteine and polyphenylalanine affinity tails', *Anal. Biochem.*, **172**, 330–337 (1988).

[22] E. Hochuli, W. Bannwarth, H. Dobeli, R. Gentz and D. Stuber, 'Genetic approach to facilitate purification of recombinant proteins with a novel metal chelate adsorbent', *Bio/Technology*, **6**, 1321/1325 (1988).

[23] T. P. Hopp, K. S. Prickett, V. L. Price, R. L. Libby, C. J. March, P. T. Ceretti, D. L. Urdal and P. J. Conlon, 'A short polypeptide marker sequence useful for recombinant protein identification and purification', *Bio/Technology*, **6**, 1204–1210 (1988).
[24] T. Moks, L. Abrahamsen, E. Holmgren, M. Bilich, A. Olsson, M. Uhlen, G. Pohl, C. Sterky, H. Hultberg, S. Josephson, A. Holmgren, H. Jornvall and B. Nilsson, 'Expression of human insulin-like growth factor I in bacteria: use of optimized gene fusion vectors to facilitate protein purification', *Biochemistry*, **26**, 5239–5244 (1987).
[25] T. Moks, L. Abrahamsen, B. Osterlof, S. Josephson, M. Ostling, S.-O. Enfors, I. Persson, B. Nilsson and M. Uhlen, 'Large-scale affinity purification of human insulin-like growth factor I from culture medium of *Escherichia coli*', *Bio/Technology*, **5**, 379–382 (1987).
[26] P. Bornstein and G. Balian,'Cleavage at Asn–Gly bonds with hydroxylamine', *Methods Enzymol.*, **47**, 132–145 (1977).
[27] B. Hammarburg, P.-A. Nygren, E. Holmgren, A. Elmblad, M. Tally, U. Hellman, T. Moks and M. Uhlen, 'Dual affinity fusion approach and its use to express recombinant human insulin-like growth factor II', *Proc. Natl. Acad., Sci. USA*, **86**, 4367–4371 (1989).

12

Protein purification: a new approach to affinity chromatography

Ken Jones
Affinity Chromatography Ltd, Ballasalla, Isle of Man

INTRODUCTION

The technique of affinity chromatogoraphy is not a mainstream chromatographic method, and yet it encompasses one of the most potentially important chromatography applications of the future — the separation of biological macromolecules. Many proteins are separated analytically in the reversed phase, but these methods are not suitable for scale-up. The problem is summarized in one word: silica. Even when bonded with strongly hydrophobic groups, silica still exhibits a degree of solubility in aqueous systems. When the pH exceeds 9, solubility becomes severe. In contrast, the use of strong alkali is much favoured by the biochemist. Caustic soda is a powerful de-pyrogenating reagent and is also very effective in removing impurities which strongly adhere to column packings. Impurities can very easily build up on the column, which then lowers capacity, restricts flow and eventually causes replacement of the packing. Large-scale media are essentially designed to operate in a multi-cycle environment. This usually involves at least one wash with molar concentrations of caustic soda. Under these conditions the matrix has to be very robust and dissolution prevented. Column fouling not only shortens lifetime but also effects the final purity of the target protein. Silica is clearly a least-favoured matrix and alternatives have to be sought when scale-up is considered.

Long before chromatography became *the* major separations technology when gas chromatography was introduced in 1952, biochemists used adsorption chromatography columns for the separation of natural products. Of the many variations of column chromatography used today, affinity chromatography as a separate technique has only been recognized in the last two decades. Despite its short history it is already accepted as a major technology at both small and large scale. It now appears to be destined to become *the* major method for protein separation and purification. It is a highly specific technique and is designed to extract a single protein species from very dilute solutions containing a multiplicity of similarly structured proteins.

Ch. 12] Protein purification: a new approach to affinity chromatography

SPECIFIC VS. NON-SPECIFIC ADSORPTION

The common feature of biological entities is their ability to recognize and bind to other molecules, often in a highly specific manner. It is this binding ability which allows all proteins to be separated and purified. The basic intermolecular forces which effect separations are electrostatic, hydrophobic, hydrogen bonding and van de Waals interactions. To these basic physico-chemical characteristics can be added size-exclusion techniques. In combination they provide a full panoply of separation methods. In general ion-exchange materials use only one of the binding forces; they are classified as non-specific reagents. When more than one binding force is used simultaneously, materials are classified 'group-specific reagents'. Hydrophobic-interaction media fall into the latter category. The relative effectiveness of non- and group-specific reagents can be described diagramatically by a three-dimensional grid (Fig. 12.1). The crude protein is represented by various-sized

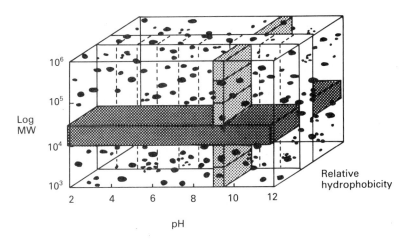

Fig. 12.1 — Effectiveness of specific and non-specific reagents.

ovoids sited within the matrix. Each ovoid represents an individual protein which has a given characteristic of hydrophobicity, size and ionic charge. If the crude mixture is passed through an ion-exchange column the proteins captured by the column by ionic charge lie within the designated block. The remaining uncaptured impurities pass through. The desired fraction is then regenerated from the column and collected as a partially purified portion. This fraction is then passed through a size-exclusion column (Fig. 12.1), the desired fraction collected as before, and then passed through a hydrophobic-interaction column. The final eluant theoretically contains only those proteins which occur at the intersection of the three blocks within the grid. Although a significant purification has taken place, the eluant is still only partially purified. Since the processes of ion exchange, size exclusion and hydrophobicity are (relatively) non-selective, the total separation can only represent a 'broad band' approach.

The attainment of high purity consequently demands further purification, and several more stages of chromatography have to be added. This applies to all sources of the protein, whether naturally occurring or produced by genetic engineering procedure. The achievement of high purity is made more difficult by the target protein inevitably existing in the crude mother liquor in co-existence with many other proteins of very similar structure. Even where only modest levels of purification are required, 5–10 separating stages can be required. For high-purity therapeutic proteins, 10–20 purification steps are not uncommon. Such lengthy multi-step operations incur high processing costs and yield reduction at each step. The replacement of any or all of the stages by a single bioselective step is a very worthwhile objective; major economic gains then result.

Fig. 12.2 compares the two approaches. The purification of the enzyme required seven purification steps by the conventional route. Initial activity in the crude liquor was 27 units, rising to 520 as the process proceeded, but yield falls to 9%. Replacement of the multi-step process by one affinity step raises yield to 71% and activity to 590 units. It has been reported that 50–80% of the production costs of a therapeutic protein occurs at the purification stage [1]. This example demonstrates why such high costs occur.

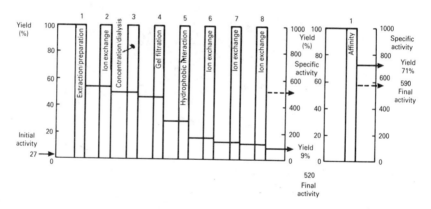

Fig. 12.2 — Comparison of multistep vs. affinity purification.

Once the technique of bioselectivity had been conceived, it was rapidly developed into a major purification routine, and materials sought which would resolve specific separation problems. Being wholly familiar with the kinetics and operating mechanisms of protein systems, biochemists quite naturally turned to other proteins as a means of creating the desired high-specificity separations. Such ligands, classified as 'natural' in contrast to 'synthetics' have first to be found, then attached to an inert matrix to create a chromatographic media. From these beginnings affinity chromatography was born. The mechanism of affinity separations is shown in Fig. 12.3. The ligand is bound onto the matrix, the bound matix packed into a

Ch. 12] Protein purification: a new approach to affinity chromatography

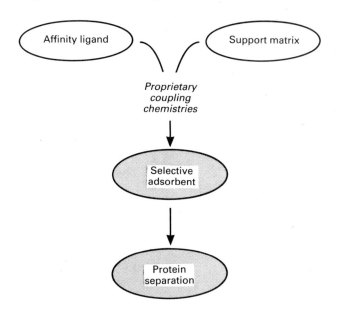

Fig. 12.3 — Mechanism of affinity separations.

column, followed by passage of the impure sample through the packing. The ligand, selected to recognize only the target protein, exclusively captures the protein whilst impurities pass through. The retained protein can be recovered from the column, usually as a single pulse, in its pure form.

Various methods of regeneration can be used, ranging from changes of pH and salt concentration, through the displacement by co-factors.

However, severe problems were still apparent with the technique. Finding suitable ligands, which tend to be rare in nature, often involved a great deal of research. It was this difficulty and expense which proved to be a major constraint in extending the horizons of affinity chromatography. For analytical separations these limitations are not too serious. Purifying small amounts of natural ligands, to operate under non-rigorous conditions, and where the cost of packing is only a small proportion of the final column price can be tolerated. At the larger scale media costs represents a much higher proportion of operational costs, and industrial quality assurance standards have to be achieved before a product can be commercially launched. There are also other impediments. Proteins tend to be unstable when operating under process chromatographic conditions and large-scale processes are essentially multi-cycle. These involve harsh regeneration steps, conditions not conducive to stability of the biological ligands. Such factors lead to short column lifetimes. This combination of probems accelerated the search for suitable, inexpensive ligands which would offer much greater stability.

Although these limitations restrained the early adoption of natural matrices at the process scale, there were unique benefits which more than offset the disadvantages. Of these, the fact that affinity ligands uniquely concentrate the target protein from very dilute streams [2], can stabilize the adsorbed protein whilst it is bound onto the column [3], and finally release it in virtually quantitative yield, proved to be of major purport. These benefits are supported by examination of Fig. 12.4 [4] where

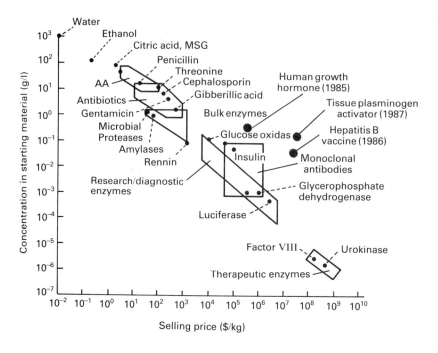

Fig. 12.4 — Product value vs. concentration.

the value of a product is expressed as a function of concentration in the mother liquor. The more dilute the solution, the greater the degree of difficulty of purification, and the higher the product value. A technology which can maximize the yield of very high value products, when the only sources are very dilute solutions, obviously has a major future.

THE MATRIX

The development of high-performance affinity liquid chromatography (HPALC) is restricted owing to the lack of an appropriate matrix which parallels the role of silica in HPLC. Alternative matrices must be capable of sustaining high pressure, whilst resisting alkali as well as exhibiting non-specific binding characteristics. It must also be of modest price for scale-up purposes. Several synthetic polymers, based on styrene–divinyl benzenes co-polymers, hydroxy ethyl methacrylates and even PTFE have been made available. Although these meet basic physico-chemical characteristics, none of the beads meet the price criterion. Until the price of synthetic matrices approaches the cost of the widely used soft gels (mainly agarose in its various forms), then agaroses are likely to maintain their current very dominant commercial position.

When processing proteins, yield maximization is of major importance. This relegates the demand for high column efficiencies (in terms of theoretical plates) to a more minor role. Equally, since proteinaceous materials are usually unstable and denature easily if subjected to harsh environments, bioselective ligands have to be stable, in turn conferring a high degree of stability onto the adsorbed target protein once bound onto the ligand. The matrix must also have very large pores to allow easy passage of bulky molecules, but having large pores implies a low surface area. This factor can only provide limited opportunity onto which high ligand concentrations can be attached, suggesting that high capacities (rate of production per unit time) are difficult to achieve. Fortunately, unlike most forms of chromatography, low ligand concentrations do not limit the attainment of high capacites. Capacities of up to 40 mg per ml of bed volume can be achieved. Speed of separation, a component which significantly effects economics, could be achieved if high-performance matrices were not restricted by the price structure. Agaraose in many ways is an ideal matrix, but its poor pressure performance will eventually drive manufacturers to find improved replacements. Despite this limitation, for large-scale applications agarose seems set to remain the favoured choice for a long time to come.

NATURAL VS. SYNTHETIC AFFINITY LIGANDS

Many elegant separations have been achieved with natural ligands, creating an aesthetic appeal which has assisted in the development of the method. Protein A has become *the* major separation technology for IgGs. Although the cost of Protein A media has recently fallen dramatically since the protecting patents expired, it is still prohibitively expensive for large-scale use. It is not merely the expense which deters its use. As a biologically derived ligand it is relatively unstable, and is prone to breakdown when used in multi-cycle processing. Regulatory authorities (e.g. FDA) are reluctant to accept inherently unstable ligand chemistries, and approvals are not given willingly. Natural ligands were, however, not the first ligands to be discovered for affinity chromatography. Synthesized ligands, with their much smaller-designed molecules, offer much higher degrees of stability and are also much less costly.

Metal ions have proved useful where the protein of interest can chelate, and amino acids and their derivatives (e.g. lysine, arginine and benzamidine) have been widely used, the latter primarily used for proteases. Of the two other general categories, hydrophobic interaction materials and those based on textile dyes have the widest applicability.

Hydrophobic interaction

Conventional reversed-phase materials with silica as the matrix are very widely employed for the analytical evaluation of protein mixtures. However, potential contamination of the product by dissolved silica, the ligand attached to the silica, and the silane residues used during bonding, all disfavour large-scale use. Nonetheless, when hydrophobic groups are bonded to the more chemically robust agaroses, they have become one of the most widely used affinity media today. Hydrophobic-iteraction (HI) media are (at best) group specific. This lack of bioselectivity results in HI reagents being used in combination with other non- or group-specific media (Fig. 12.2).

Textile dye ligands
Over 1000 papers have been published on the use of textile dyes for protein purification. The vast range of proteins purified include blood proteins, dehydrogenases, kinases, oxidases, proteases, nucleases, transferases and ligases. Of the many dyes used, only two have become widely established commercially. These were (originally) Cibacron Blue F3G-A™, and Procion Red H-E3B™. The former dye was manufactured by Ciba and sold exclusively by them purely for textile dyeing. It has the structure shown in Fig. 12.5. and belongs to the C.I. Reactive Blue 2 generic

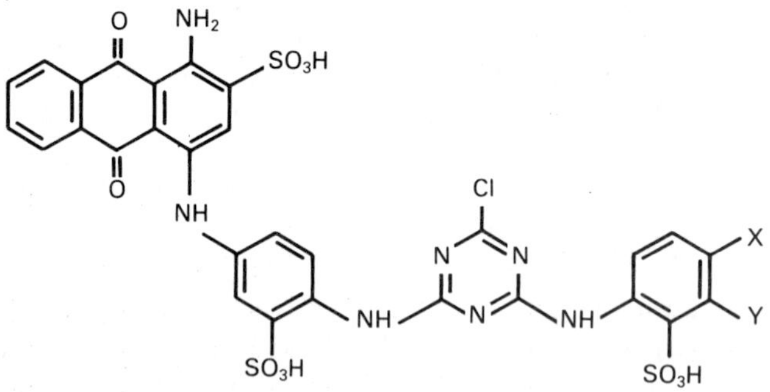

Fig. 12.5 — Cibagon Blue.

class of dyes. It has not been manufactured for many years and this has led media suppliers to offer other members of the C.I. Reactive Blue 2 class, sometimes wrongly designating these products as Cibacron Blue F3G-A. In most members of the C.I. Reactive Blue 2 class, a different isomer than Cibacron F3G-A is used. The location of the sulphonic acid group on the pendant ring is either in the *meta* or *para* position, or more usually a mixture of the isomers. The ratio of the isomers (*ortho*, *meta*, *para*) can vary significantly from manufacturer-to-manufacturer, and from batch-to-batch even from the same manufacturer. Since dyestuff manufacturers do not disclose the structures of their products, the media supplier often has no means of knowing the exact chemical structure of the product being offered. It has been demonstrated that even the most minor isomeric structural changes can profoundly effect the chromatographic performance of synthetic ligands.

Since an essential feature of all chromatographic processes is the necessity for exact repeatability from column-to-column, this is clearly one of the reasons why textile dyes have not been used more extensively at the large scale. Textile dyes are bulk chemicals which command a correspondingly low price, usually below £15/kg. To meet this price the manufacturer cannot economically, and indeed has no need to, remove many by-products which are irregularly produced at every stage of the manufacturing process. Such impurities are undoubtedly detrimental in attempting to offer any guarantee of chromatographic reproducibility. They bind onto the column at different rates and with different bonding mechanisms, thus providing an inherent instability which results in leakage, lack of reproducibility and short lifetimes.

A further factor, and of equal significance, is that the bonding process between the main structure of the dye and the matrix has been insufficiently researched. By applying published bonding methods, even with very pure dye, some dye leakage still results. All commercially offered textile-dye media leak to varying degrees, stemming either from the use of impure textile dyes and poor bonding methodology, or more usually both. The leakage rate is dependent upon the manufacturer and the matrix used as shown in Fig. 12.6. Nevertheless, in volume terms products based on

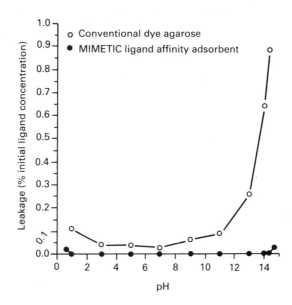

Fig. 12.6 — Comparison of ligand leakage from MIMETIC ligand A6XL and conventional textile dye agarose. Data determined from 1.0 g adsorbent samples shaken in 5 ml of test solution for 48 h at 4°C. Assay detection limit: 0.02% of initial immobilized ligand concentration, equivalent to 0.5 nmol per g moist gel.

C.I. Reactive Blue 2 class dyes are by far and away the largest-selling affinity ligand. Their usefulness significantly outweighs their major disadvantages. As a basic observation this statement begs the question; if such ligands could be made leaktight, and the range extended, how much more useful would they be? With this as an objective, a new approach has been devised.

MIMETIC™ ligands

In work led by Dr C. Lowe, latterly at the Institute of Biotechnology, University of Cambridge, UK, it was initially demonstrated that ligands must be of the highest chemical purity if leakage was to be reduced and reproducibility achieved. Secondly, isomeric purity was necessary to build upon the first-stage improvements, followed by the development of new bonding technologies if leakage was to be finally eliminated. The fourth step was to investigate what effect structural changes, if any, would have on enhancing the efficacy of the basic dyes structures. Lastly, could broad-spectrum ligands be designed, where a minimum number of ligands would separate a maximum number of proteins?

The last two stages were significantly assisted by the availability of computer-aided molecular-modelling facilities. By using Evans and Sutherland, and Brookhaven protein data bases, in combination with Sybil™ and Macromodel™ programs, it proved possible to develop new insights into protein-synthetic ligand interactions. One of the major limitations in the use of modelling design concepts is that only some 400 protein structures are available on the Brookhaven data base. Protein structures are deduced from X-ray crystallography data, and proteins only crystallize when of extreme purity. Fortunately capillary zone electrophoresis (CZE), capable of developing up to 250 000 theoretical plates in 15 minutes, promises to be a major aid in determining the purity of proteins. When run in tandem with affinity techniques, the production of crystalline proteins should now accelerate. This will result in an extended database, from which new ligands can be designed, enabling more efficient purification, and so around the cycle.

One of the advantages of affinity chromatography is that an infinite range of designed structures, capable of bioselectively recognizing any protein from whatever source, is possible. Even so, this realization has to be tempered by the commercial undesirability of having a very large number of ligands, each of which can only make a marginal contribution to the whole separations area. Key binding centres associated with generic groups of proteins should ideally be identified. From these data universal ligands can be designed which will separate a large range of protein types. Application of this philosophy has resulted in a range of MIMETIC™ products. The structural changes of this made within this range are so extensive that they no longer have much resemblance to textile dyes. More correctly they are ligands which are incidentally coloured rather than coloured by design. The difficulty of achieving virtually zero leakage was further highlighted during development of quality assurance procedures, essential for a commercial launch of the products. Although leakage had been largely eliminated when compared to textile dyes, it became apparent that it was necessary to design bonding processes into the basic chemical structure of the new ligands. The success of this approach can be judged from the results given in Fig. 12.6.

Each of the agarose-based MIMETIC products can be depyrogenated with 2 M NaOH, and withstand treatments with strong chaotropes and detergents, as well as 10 mM HCl over many cycles. These robust materials have been recycled through extremes of pH over 800 times without apparent deterioration. By developing the proprietary bonding methods even further, it also proved possible to increase the physical strength of the agarose beads. Consequently considerably higher flow rates

Ch. 12]   **Protein purification: a new approach to affinity chromatography**   245

can be achieved when compared to conventional agaroses (Fig 12.7) adding more substance to the economic benefits previously outlined.

METHOD DEVELOPMENT TO THE LARGE SCALE

Gravity or very-low-pressure chromatography separations are second nature to biochemists, all the way from analytical columns up to columns with volumes exceeding 1000 litres. At some stage all protein purification involves the use of packed columns, whether ion exchangers, size exclusion, chelating systems, hydrophobic interaction, or affinity ligands. Non- and group-specific separation strategies in combination have historically been adopted, and quite naturally biochemists tend to turn to these well-tried methods. Any criticsm of this conventional approach should be suppressed. The problems faced by biochemists are rarely appreciated by those chromatographers more familiar with HPLC. Apart from having to develop a new assay method with each new protein, at best very little knowledge, if any, on the structure of the protein of interest is known. What little data that are known are often restricted to the basics of molecular weight and isoelectric point. Almost always an empirical approach has to be adopted, and development of a separations routine owes little to logic. This all too often results in a great deal of tedious and time-consuming work, and which usually results in an inefficient process involving several stages of separation, as shown in Fig. 12.2. Although highly specific methods such as affinity chromatography would obviously be preferred, the time involved in exploring the field for the one highly specific natural ligand which will effect the separation is a very daunting task. Now, however, the availability of the inexpensive MIMETIC range of media, with their high purity extreme stability and remarkable ability to separate a very wide range of protein types, offers an attractive alternative route. Furthermore, once a separation has been determined the basic structure of the ligand can often be re-designed by means of computer-assisted molecular modelling, enhancing specific characteristics such as capacity and/or specificity.

As a convenience to the researcher, ten selected broad-based agarose-bonded MIMETIC ligands have been collected into a PIKSI™ module (Protein Identification Kit by Sorbent Identification, Fig. 12.8). Of a randomly selected and very wide range of proteins so far analysed most give some degree of separation on one or more ligands. Furthermore, over 50% of proteins tested gave a significant separation. Analysis time is 30–45 minutes. A profile of the separation is then generated (Fig. 12.9), indicating the specific ligand which offers the most effective separation. Haemoglobin was used in the example shown. MIMETIC Green 1 gave a 97% adsorption, with a 95% recovery (not shown). Optimization can then follow on the selected media, scaling if appropriate to the development scale.

Scale-up of affinity media is very easy. When coupled to the unique ability of being able to process very dilute solutions, it suggests that direct scale-up from laboratory to hundreds of litres is possible. Where product value is very high (Fig. 12.4), the use of expensive matrices may be justified. MIMETIC ligands can be bonded onto most high-performance matrices with the same degree of stability exhibited with agarose matrices. When product values exceed £100 000 per kg, media

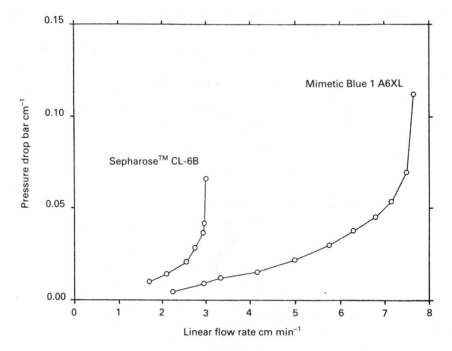

Fig. 12.7 — Comparison of pressure drop vs. flow curves for conventional 6% cross-linked agarose and MIMETIC ligand A6XL affinity adsorbents. Data determined from 14 cm columns of gel, 3.2 cm in diameter.

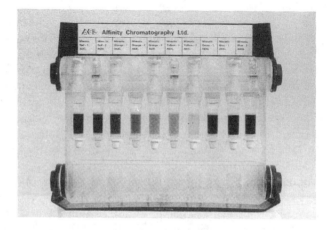

Fig. 12.8 — PIKSI™ module.

Ch. 12] Protein purification: a new approach to affinity chromatography 247

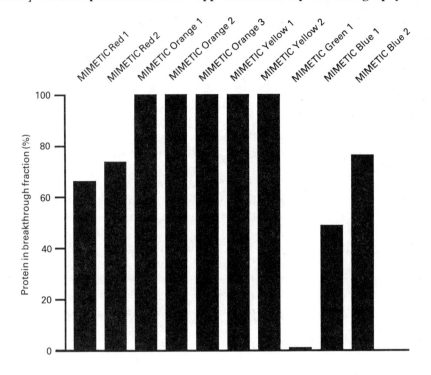

Fig. 12.9 — Profile of the separation.

costs may become less important, but again only when a large number of cycles can be completed before replacement has to be considered. As demand for affinity media increases, and larger volumes are manufactured, so price will be driven down. At this juncture the interests of analysts and large-scale users converge, and direct scale-up from HPLC analytical columns to large-scale HPALC will be commonplace.

EXAMPLES OF DESIGN

Manipulation of protein models on the Brookhaven data base (or simulations thereof) can expose those elements which may be binding centres, and onto which can be juxtaposed appropriate groups designed into a synthetic ligand backbone. One example is a ligand capable of separating kallikrein from trypsin. Trypsin cleaves polypeptides adjacent to organic residues, whilst the more selective kallikrein has a secondary binding site requirement for phenylalanine. Arginine is known to be a primary binding site. A new compound was developed which incorporated both benzamidine and an unsubstituted phenyl ring. This ligand was successfully synthesized [5]. Upon immobilization the ligand displayed high affinity for kallikrein but did not bind trypsin, an enzyme which normally binds to benzamidine agarose and is a common contaminant or kallikrein preparations [6]. A similar approach was used in the use of a phosphonic acid group, a known inhibitor of alkaline phosphatase. This was incorporated into the terminal aminobenzene ring of Reactive Blue 2, providing a 330-fold purification of calf intestinal phosphatase from crude intestinal extract in one step [7].

The elongation of a dye structure exemplifies another approach. C.I. Reactive Blue 2 binds the enzyme horse-liver alcohol dehydrogenase [8]. This enzyme is known to bind the dye at the same site as the natural co-factor NAD^+. X-ray crystallography revealed that the anthraquinone ring, benzenesulphonate bridging ring and triazine ring of the dye (Fig. 12.5) were bound in similar positions to the adenine, adenyl ribose and pyrophosphate groups of NAD^+. However, the dye was not an exact mimic of NAD^+ since the terminal aminobenzene sulphonate ring of the dye is located in a position some 10 Ångströms from the nicotinamide binding site [9]. Computer modelling showed that the dye was insufficiently long, and lacked the necessary flexibility to exactly mimic the conformation of enzyme-bound NAD^+. Adding an aminoethyl spacer group in the centre of the molecule allowed the new structure to adopt a very similar conformation to bound NAD^+. Minor changes in other groups were also made. The new biomimetic structure was synthesized and was shown to resolve different species of horse-liver alcohol dehydrogenase, whilst affording a high degree of enzyme purifaction (64-fold) from a crude horse-liver extract [10].

CONCLUSIONS

Affinity separations using synthesized ligands are the most-favoured route for many large-scale protein separations. Of the natural ligands, only Proteins A and G are likely to retain their dominant position, primarily because of their uniqueness in purifying UgGs. These will eventually be replaced as more accurately designed synthesized ligands become available. The relatively simple process of making structural changes to already effective existing synthetic molecules is of the greatest possible interest to biochemists, but the need for pyrogen-free systems is a complicating factor. It excludes as a starting point any packing based on silica. Additionally, since reversed-phase HPLC is a low-specificity method, there is limited mileage in considering such an approach for scale-up purposes. The ability to engineer specific characteristics into the ligand appears to be destined to make synthetic ligands the major tool for the large-scale user.

REFERENCES

[1] N. E. Pfund and K. G. Charles, *'The Wheat from the Chaff'*. Hambrecht and Quist, San Francisco (1987).
[2] P. Reyes and R. B. Sunquist, *Anal. Biochem.* **88**, 522 (1978).
[3] D. McCormack, *Biotechnology*, **6**, 158 (1988).
[4] P. Knight, *'Downstream Processing'*. *Biotechnology*, **7**, 777 (1989).
[5] M. Johnson, J. P. Overington and A. Sali, *'Current Research in Protein Chemistry'*. Ed by J. Villafranca, Academic Press, New York (1990).

[6] N. C. Burton, PhD Thesis, University of Cambridge, UK (1990).
[7] N. M. Lindner, R. Jeffcoat and C. R. Lowe, *J. Chromatogr.*, **473**, 227 (1988).
[8] S. J. Burton, C. V. Stead and C. R. Lowe, *J. Chromatogr.* **455**,
[9] J.-F. Biellemann, J.-P. Samana, C.-J. Branden and H. Eklund, *Eur. J. Biochem.*, **102**, 107 (1979).
[10] S. J. Burton, PhD Thesis, University of Cambridge, UK (1988). Registered trade marks are PIKSI and MIMETIC to Affinity Chromatography Ltd., Cibacron to Ciba-Geigy, Procion to I.C.I. Ltd.

13

An introduction to large-scale enantioseparation

Charles A. White
Fisons Plc, Loughborough, UK

INTRODUCTION

Before a discussion on preparative enantiomeric separation can take place a basic discussion on what chirality is, why it is important and how analytical-scale separations are performed must take place because these parameters are frequently overlooked in the basic introductory texts on HPLC. A reader who has a basic knowledge of HPLC may well understand the mechanisms involved in reversed-phase, normal phase and ion exchange whilst gel permeation and affinity chromatography may well be understood to a very limited extent. Without a basic understanding of chirality, the mechanisms involved in enantioseparations are difficult to comprehend and discussion on preparative separations pointless.

What is chiralty?
A molecule is called chiral (from the Greek word, χειρ, meaning hand) if it is not superimposable with its mirror image (Fig. 13.1). The two different non-superimposable forms of a chiral compound are named optical isomers or enantiomers whilst a mixture of the two forms is known as a racemic mixture.

In Fig. 13.1 the two forms of butan-2-ol are not superimposable but behave in an

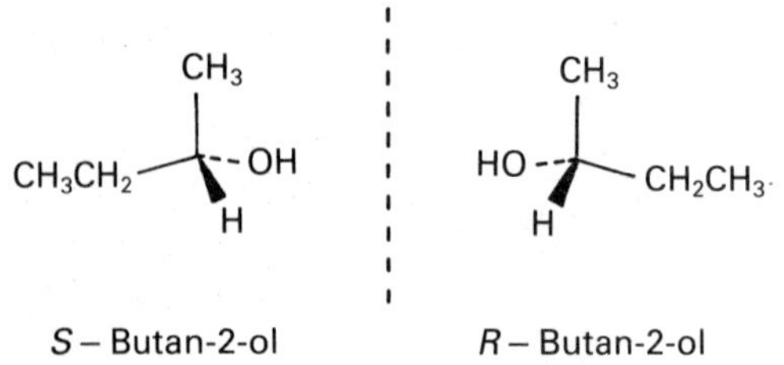

Fig. 13.1 — Stereochemical structure of butan-2-ol.

An introduction to large-scale enantioseparation

identical manner in most situations. However, when put in a chiral environment the differences become evident. A group of men may look identical, each having two arms, two legs, one head, etc., and behave in a similar manner but when it comes to writing, a number will use their left hand. In a chiral environment they behave differently.

What causes chirality?
As shown in Fig. 13.1, a carbon atom which has four different substituents produces a chiral centre. What is often not recognized is that other atoms can also produce chiral centres. Fig. 13.2 shows how sulphur, nitrogen, phosphorus and boron can also

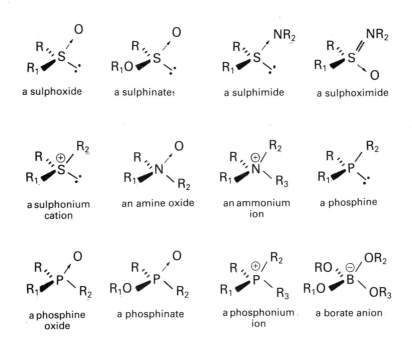

Fig. 13.2 — Examples of stable chiral molecules with chiral atoms other than carbon.

produce stable chiral structures. In addition chiral axes and planes can be obtained. Examples of chiral axes (Fig. 13.3) include not only rigid structures such a dialkenes (Fig. 13.3(a)) but also structures in which rotation about a single bond is restricted owing to steric interactions (Fig. 13.3(b)). Similarly chiral planes (Fig. 13.4) can be produced by rigidly fixing two atoms as in the substituted phsophoric acid (Fig. 13.4(a)) or as a result of steric interactions forcing distortions such as left- and right-handed spirals in helicenes (Fig. 13.4(b)). Organometallic complexes also generate chiral molecules as a result of bonding between the metal ion and organic compounds as shown in Fig. 13.5. Rotation about the double bond or planar axis is restricted as a

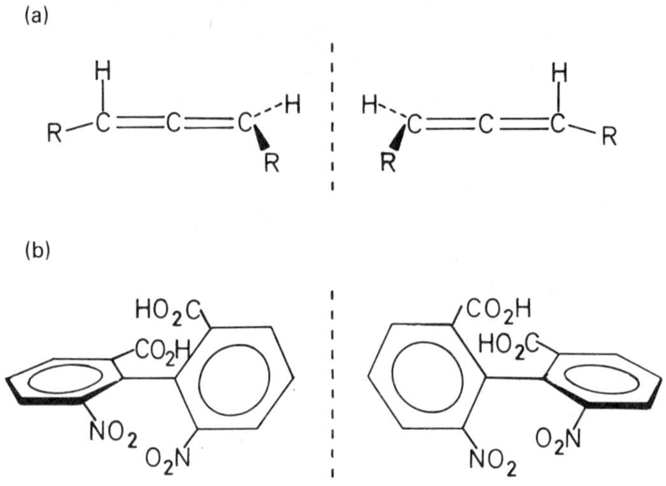

(a)

(b)

Fig. 13.3 — Examples of stable chiral molecules with a chiral axis.

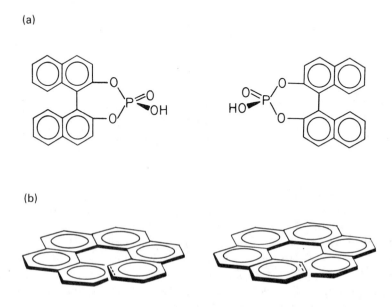

Fig. 13.4 — Examples of stable chiral molecules with a chiral plane.

Ch. 13] An introduction to large-scale enantioseparation

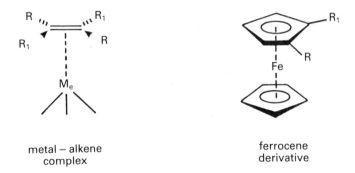

Fig. 13.5 — Examples of chirality generated in organometallic complexes.

result of metal complexation, thereby allowing generation of stable optical isomers. Therefore chirality extends beyond the simple chemistry of carbon into many chemical types, some of which may not initially be simple to recognize.

Why is chirality important?

Nature, in order to simplify the vast number of synthetic possibilities, relies on chirality. Biosynthesis and metabolism are heavily controlled by chirality. For example production of glucose results in a single compound (D-glucose) being metabolized rather than a mixture of the sixteen possible hexoses. Similarly only the L-amino acids are produced, not a mixture of D- and L- isomers. With enzymes frequently only able to recognize, bind and react with a specific optical isomer, it is not surprising that chiral products involved in biological reactions behave differently. What is surprising is the extent to which chiral products affect our lives and the industries which are involved.

INDUSTRIES WHICH REQUIRE ENANTIOSEPARATIONS

Pharmaceutical industry

The production of drugs was severely affected by the thalidomide tragedy in the early 1960s. Had the knowledge of chirality been fully understood and methods known for production and analysis of specific isomers, then the tragedy might not have happened. Nowadays much more is known about the chirality and how different isomers react in the body [1]. It is not only that one isomer reacts and the other does not but in some instances different isomers can have different effects (Table 13.1).

A report produced in 1982 [2] showed that of 1675 drugs produced only 28% were natural or semisynthetic. Of those, 98% were chiral and were marketed as single isomers. Of the 72% of the drugs that were synthetic 40% were known to be chiral but the majority (88%) were sold as racemates or mixed isomers. Today, with the impending regulation on production and use of chiral drugs, the need to monitor production, report isomeric composition of products and study pharmacological

Table 13.1 — Examples of pharmaceutical products which show the effect of chirality

Compared	Isomer	Effect
Thalidomide	S-Isomer	Teratogenic
	R-Isomer	Sleep inducing, anti-nausea
Barbiturates	S-Isomer	Depressant
	R-Isomer	Convulsant
Opiates	R, S-Isomer	Narcotics
	S, R-Isomer	Non-addictive cough mixture
Labetalol	S, R-Isomer	Alpha-blocker
	R,R-Isomer	Beta-blocker
Penicillamine	D-Isomer	Anti-arthritic
	L-Isomer	Toxic

effects of drugs has resulted in 70% of the chiral chromatography market being accounted for by the pharmaceutical industry with emphasis on both analytical and small preparative-scale processes being made.

Agrochemical industry
A similar situation to the pharmaceutical industry exists in the agrochemical industry although the actual level of the chiral chromatography market is only about 25%. The need to produce the most-effective pesticide economically and with the minimum of toxic effects is driving the industry to produce single isomers where possible. A report produced in 1981 [3] showed that of 550 pesticides 98% were synthetic products of which only 17% were chiral materials but less than 8% of these were marketed as single isomers.

Food and drink industry
This industry is becoming more involved with chiral separation as a result of the differences between chemical and biochemical processes producing different isomeric compositions. Typical examples include conttrol of fermentation processes and storage effects when a single isomer is converted to a racemic mixture as time proceeds. Two topical examples are found in artificial sweetener and olive oil production. The S-isomer of asparagine is bitter whilst the R-isomer is sweet and so chiral analysis of both the precursor to the sweetener, Aspartame, and of the product during storage is essential to ensure that the sweetener and products containing it are not affected by adverse taste effects. Olive oil is a typical example of products which are monitored to prevent adulteration of natural products by cheap chemically produced materials.

Petrochemical industry
Many of the products from the petrochemical industry are important precursors for synthetic processes. With the importance of stereospecific synthesis, the need to monitor production of chiral precursors (halogenated hydrocarbons, polyalkenes, etc.) is becoming important.

Another interesting area for chiral analysis is in the fingerprinting of samples during exploration since different oil fields are produced from different materials at different times and therefore ratios of stereoisomers can be found in samples from adjacent fields.

ANALYTICAL CHIRAL SEPARATIONS

Classical methods of optical resolution, such as recrystallization of diastereomeric salts, can readily be performed on the large scale but such methods are rarely suitable for industrial scale-up. For this reason it is easy to appreciate the advantages of chromatographic separations for large-scale enantioseparations.

The first commercially available stationary phase for chiral HPLC was introduced in 1981 [4] and development has continued in the past ten years such that over 50 different stationary phases are now commercially available. In an attempt to provide the analyst with assistance in the development of an assay and selection of which stationary phase to use, a classification of the basic types has been proposed [5]. This classification is based on how the complexes between solution and stationary phase are formed (Table 13.2). It serves as a useful basis for reviewing those stationary phases which are suitable for large-scale enantioseparations.

Table 13.2 — Classification of chiral stationary phases

Classification	Interaction
Type I	Attractive interactions, hydrogen bonding, π–π interactions and dipole interactions
Type II	Both attractive interactions and inclusion complex formation
Type III	Retention via formation of inclusion complexes within chiral cavities
Type IV	Ligand exchange mechanisms with metal complexes
Type V	Hydrophobic and polar interactions with bound protein phases

Type I chiral stationary phases

This type of stationary phase is the most widely researched and contains the largest number of commercially available products. In general terms a chiral moiety is bonded, via a non-chiral spacer, to silica. Complex formation between solute molecules and the stationary phasees involves a combination of hydrogen bonding, dipole/dipole interaction, interactions and steric hindrance (Fig. 13.6). Dalgliesh [6] proposed a three-point interaction model for chiral recognition on Type I stationary phases [7].

The major contribution to stationary phases of this type was made by Pirkle and his co-workers [8]. Their initial stationary phases were based on dinitrobenzoyl

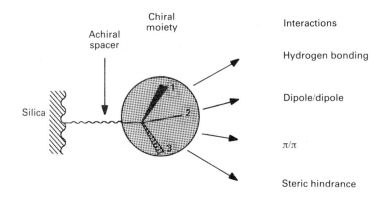

Fig. 13.6 — Type I chiral stationary phases and the type of interaction which can be involved in chiral separations.

derivatives of amino acids, particularly phenylglucine and leucine, coated onto silica materials. These so-called ionic phases showed excellent efficiency but were limited to low-polarity solvent systems if the chiral phase was to be preserved. The introduction of covalently bound derivatives two to three years later allowed a much wider range of solvent polarity to be used thereby greatly extending the range of materials which could be separated. These π-acceptor phases were followed by a family of π-donating phases based on β-naphthoyl derivatives of alanine. This family has now been extended to include a range of diverse structural types frequently possessing two chiral centres and amide or urea moieties. One of the drawbacks of this second type of Pirkle-concept stationary phase is the frequent need for derivatization (using π-doonating groups such as dinitrobenzoyl, dinitrophenylurea or dinitrophenylcarbamate derivatives) although, in some cases, the increase in detectability can offset the disadvantages of derivatization.

Table 13.3 lists the more common Type I stationary phases which are commercially available.

Type II chiral stationary phases
Derivatives of cellulose and, more recently, amylose have been found to have the ability of chiral recognition. Whilst the mechanism of interaction has not been fully understood, the inclusion of solutes in cavities of the polymer structures via attractive forces including hydrogen bonding, dipole/dipole interactions and π/π interactions. The different substituents are thought to change the ability of hydrogen bond formation [9]. Many of these types of derivatives, bonded onto silica, are commercially available (Table 13.4).

Type III chiral stationary phases
This is a diverse group of stationary phases, united only by their mechanism of action, namely inclusion of the solute within a chiral cavity. The group encompasses carbohydrate polymers, crown ethers and synthetic polymers (Table 13.5)

Table 13.3 — Structures of the type I chiral stationary phases

(R)-N-(3,5-Dinitrobenzoyl)phenylglycine [covalent]
(R)-N-(3,5-Dinitrobenzoyl)phenylglycine [ionic]
(S)-N-(3,5-Dinitrobenzoyl)phenylglycine [covalent]
(S)-N-(3,5-Dinitrobenzoyl)leucine [covalent]
(S)-N-(3,5-Dinitrobenzoyl)leucine [ionic]
(R)-N-(β-Naphthoyl)alanine [covalent]
(S)-N-(β-Naphthoyl)alanine [covalent]
(S)-1-(α-Naphthyl)ethylaminoterephthalic acid [covalent]
(S)-2-(4-Chlorophenyl)isovaleroyl-(R)-phenylglycine [ionic]
(1R,3R)-trans-Chrysanthemoyl-(R)-phenylglycine [covalent]
N-tert-Butylaminocarbonyl-(S)-valine [covalent]
N-3,5-Dinitophenylaminocarbonyl-(S)-valine [covalent]
(S)-N-1-(α-Naphthyl)ethylaminocarbonyl-(S)-valine [covalent]
(R)-N-1-(α-Naphthyl)ethylaminocarbonyl-(S)-valine [covalent]
(S)-N-1-(α-Naphthyl)ethylaminocarbonyl-(R)-phenylglycine [ionic]
(R)-N-1-(α-Naphthyl)ethylaminocarbonyl-(R)-phenylglycine [ionic]
(R)-α-Methylbenzylurea [covalent]
(R)-(+)-Naphthylethylamine polymer

Table 13.4 — Structures of the type II chiral stationary phases

Cellulose triacetate
Cellulose tribenzoate
Cellulose trisphenylcarbamate
Cellulose tricinnamate
Cellulose tris(3,5-dimethylphenylcarbamate)
Cellulose tris(p-chlorophenylcarbamate)
Cellulose tris(p-methylphenylcarbamate)
Cellulose tris(p-methylbenzoate)
Amylose tris(3,5-dimethylphenylcarbamate)
Amylose tris(S-1-phenylethylcarbamate)

Microcrystalline cellulose triacetate
This cellulose derivative, produced by the heterogenous acetylation of microcrystalline cellulose, owes its chiral recognition to the crystalline cellulose, owes its chiral recognition to the crystalline structure of the starting material which must be preserved during derivatization [10]. It has different chiral and structural selectivities to the cellulose triacetate bound to silica (the type II phase).

Table 13.5 — Structures of the type III chiral stationary phases

Microcrystalline cellulose triacetate
β-Cyclodextrin
α-Cyclodextrin
18-Crown-6 chiral crown ether
Poly(triphenyl methyl methacrylate)
Polyacrylamide
Polymethacrylamide

Cyclodextrins
Cyclodextrins are a family of cyclic D-gluco-oligosaccharides produced from starch by enzymic action. Whilst cyclodextrins containing between six and twelve D-glucose units can be produced, only those containing six, seven or eight residues are commercially available. These cylindrical molecules (Fig. 13.7) have a hydrophobic cavity and a hydrophilic rim. This rim contains between 12 and 18 secondary hydroxyl groups (depending on the number of D-glucose residues in the cyclic structure). The size of the hydrophobic cavity is such that the α-cyclodextrin (six

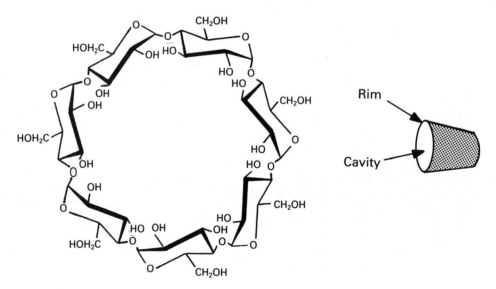

Fig. 13.7 — The structure of β-cyclodextrin and its conformation in aqueous solution.

residues in the ring) can accommodate a single phenyl ring; whilst the β-cyclodextrin (seven residues) and γ-cyclodextrin (eight residues) can accommodate substituted single- and multiple-ring systems.

Initially the cyclodextrins were cross-linked to give a gel structure but now high-efficiency columns are prepared by attaching the cyclodextrin to silica via the primary hydroxyl groups [11]. More recently derivatives of the cyclodextrins produced by reacting some or all of the secondary hydroxyls [12] have been produced to alter the chiral specificity. The commercially available derivatives include acetyl ether, hydroxypropyl ether, naphthylethyl carbamates and *para* -toluoyl ether.

Crown ethers
The synthetic macrocyclic polyethers known as crown ethers can form selective complexes with suitable cations. The 18-crown-6 crown ether can incorporate the ammonium ion via ion-dipole interactions with the oxygen atoms (Fig. 13.8).

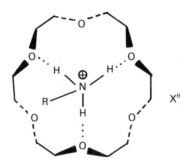

Fig. 13.8 — The structure of the complex formed by 18-crown-6 and a protonated amino compound.

Production of optically active crown ethers, chemically bonded to silica or a polymer support [13] has led to chromatographic supports capable of separating amino acids and other amino compounds.

Synthetic polymers
Many synthetic polymers, particularly those of the polyvinyl family, exist in the solid state as left-hand and right-handed helices. In solution, the high mobility of the polymeric chain prevents optical activity being shown. However, if large groups are introduced it is possible to restrict mobility and maintain a chiral helical structure in solution. Such polymers, used as bulk polymers or as lower molecular weight polymers adsorbed on a silica support [14] can be used for chiral separations with solute molecules being included in the chiral cavities within the polymer through a variety of mechanisms including hydrogen bonding.

Type IV chiral stationary phases
Chiral ligand-exchange chromatography has been recognized since 1970 but the limited range of solute molecules which can be separated has restricted the development of this method. An amino acid, more commonly proline, hydrocyproline or

valine, is attached to silica or a polymeric support. An eluent containing copper ions (less commonly nickel, cobalt, zinc or cadmium ions) produces a metal complex capable of binding amino acids and some other amino-containing compounds [15].

Type V chiral stationary phases
The complexity of protein structures produces very specific binding sites, most commonly encountered in enzymes. A number of proteins have been shown to bind specific enantiomers of a solute via a combination of electrostatic or hydrophobic interactions (Fig. 13.9). When attached to silica proteins such as bovine serum

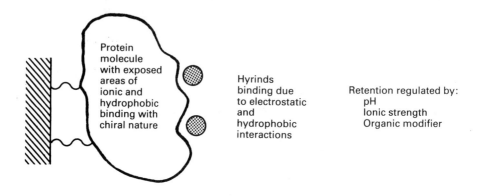

Fig. 13.9 — The type of interactions which can be involved in chiral separations using type V protein phases.

albumin [16] and α-acid glycoproteins [17] produce highly selective but low-capacity chiral stationary phases. Retention and stereoselectivity can be altered by pH, ionic strength of the usual phosphate buffer, or by organic modifiers such as propanol, dimethyloctylamine or ion-pair reagents. Recently other proteins have been proposed as alternatives to the commercially available supports. These include ovomucoid [18] and human serum albumin.

SCALE-UP OF ANALYTICAL SEPARATIONS

Scale-up analytical separations for the production of highly purified samples has several advantages, in particular the ability for direct separation using automated procedures. Conventional chromatographic processes can be scaled-up using wider pore columns to allow up to several hundred gram quantities as shown in Table 13.6. When the chiral stationary phase is fully bound to the silica no loss of valuable optically active reagents occur. In addition the optical purity of the reagent used as a stationary phase is not crucial since it only affects the separation factors obtained. Therefore, superficially, production of optically pure isomers by chromatographic processes does appear to be an economic possibility.

However, the separation factors obtained for chiral separations are frequently very low, being of the order of 1.1 to 1.5 rather than values of 5 to 10 or more which

Table 13.6 — Load cepactors of columns of different diameters

Internal diameters (nm)	Sample size
4.6	0.2–3 mg
10	5–20 mg
22/25	20–200 mg
50	1–5 g
75	1–10 g
100	4–20 g
150	5–30 g
200	10–50 g
250	15– 0 g
300	20–100 g
1000	100 g–1 kg

can be encountered in reverse-phase separations. This has the effect of limiting the amount of sample which can be loaded onto a given column size. To increase the sample size, therefore, requires a large column which increases costs and preparation time showing the trade-off which exists in the chromatographic triangle of speed resolution and capacity. Therefore, to make large-scale separations an economic possibility the following criteria must be satisfied:

    non-compressible support,
    good stability,
    broad applicability,
    acceptable cost.

Columns such as the type V protein phases, costing up to £1000 for analytical scale and with low capacities, do not lend themselves to preparative use. Semi-preparative columns of 10 mm i.d. are available for separation of about 10 mg amounts of material for laboratory analysis purposes. Type IV columns are not appropriate for preparative use since the removal of the metal ions requires subsequent separations procedures.

Type I to III phases lend themselves to large-scale works with columns up to 100 mm i.d. being readily available at prices which do not make processes uneconomic. Therefore, direct chromatographic separations of up to 10 g of materials are possible, allowing production of material for toxicity testing, etc. Probably because of the sensitivity of the work in this area there are a few examples reported in the literature.

Microcrystalline cellulose triacetate
Microcrystalline cellulose tracetate is the most readily available and cheapest phase but the low efficiency, poor reproducibility and the need for the slow flow rates owing to the limited dimensional stability tend to detract from its use. Blaschke [19] reported separation on the gram scale using 35 mm i.d. column but a 48 hour time scale was needed (Fig. 13.10),

Pirkle-type phases

The wide applicability and relatively longer availability of the Pirkle-type phases has meant that there are more reports available, ranging from simple flash chromatographic techniques [20] to single-gun prep-scale separations of 50 g of racemates. The ability to remove organic solvents used with these columns is a furtherr advantage of using these columns. Typical of the more common separations is the separation of up to 10 g quantities of material [8] using 50 mm i.d. columns (Fig. 13.11).

FUTURE DEVELOPMENTS

If the use of chiral supports is to develop into a full process-scale or into a more widely used prep-scale method, alternative approaches must be taken since, amongst other things the cost of the column is a limiting factor. The work of Vigh into displacement chromatography [21] and of Armstrong into centrifugal partition chromatography [22] indicates that these techniques allow significantly greater column loads. This allows larger scale separations on analytical or semi-prep columns without resorting to multiple injections. However, the prospects for full process-scale purification of enantiomers does not look feasible and once it has been identified that a single isomer is the required product, then the optimum route would be via enantiospecific synthesis, with chromatography being reserved for optical purity testing and lab-scale production of mg quantities for toxicity testing, etc.

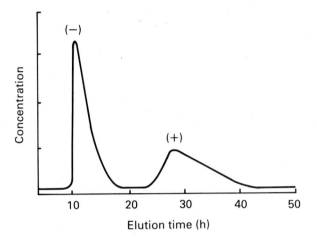

Fig. 13.10 — Separation of 2.1 g of racemic oxaapadol into enantiomers using a column (38×700 mm) packed with microcrystalline cellulose triacetate, eluent: 95% ethanol; flow rate 90 ml/h.

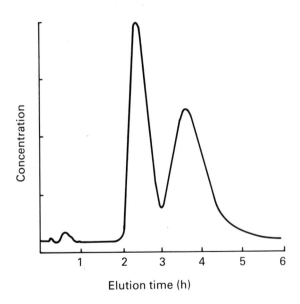

Fig. 13.11 — Separation of 1.6 g of racemic 5-(1-naphthyl)-5-(4-pentenyl) hydantoin on a Pirkle column (25×750 mm).

REFERENCES

[1] G. T. Tucker and M. S. Lennard, *Pharmaceut. Ther.* **45**, 309 (1990).
[2] K. Kleeman, U. Engel and G. Thieme, *Pharmazeutische Wirkstoffe*, Veerlag, Stuttgart (1952).
[3] E. Y. Spencer, *Guide to chemicals used in crop protection*, 7th edition Canadian Govt. Publ. Centre (1981).
[4] W. H. Pirkle, J. M. Finn, J. L. Schreiner and B. C. Hamper, *J. Am. Chem. Soc.* **103**, 3964 (1981).
[5] I. W. Wainer, *Trends Analyt. Chem.* **6**, 125 (1987).
[6] C. E. Dalgliesh, *J. Chem. Soc.* 3940 (1952).
[7] R. J. Baczuk, G. K. Landram, R. J. Dubois and H. C. Dehm, *J. Chromatogr.* **60**, 351 (1971).
[8] W. H. Pirkle and J. M. Finn, *J. Org. Chem.* **47**, 4037 (1982).
[9] Y. Okamoto, M. Kawashima and K. Hatada, *J. Chromatogr.* **363**, 173 (1986).
[10] H. Hakli, M. Mintas and A. Mannschreck, *Chem. Ber.* **112**, 2028 (1979).
[11] D. W. Armstrong and W. Demond, *J. Chromatogr. Sci.* **22**, 441 (1984).
[12] D. W. Armstrong, *J. Liquid Chromatogr.* **13**, 18 (1990).
[13] L. R. Sousa, G. D. Y. Sogah, D. H. Hoffman and D. J. Cram, *J. Am. Chem. Soc.* **100**, 4569 (1978).
[14] Y. Okamato, S. Honda, I. Okamoto, S. Murata, R. Noyori and H. Takaya *J. Am. Chem. Soc.* **103**, 6971 (1981).
[15] G. Gübitz, W. Jellenz and W. Santi, *J. Chromatogr.*, **203**, 377 (1981).
[16] S. Allenmark, B. Bongren and H. Borén, *J. Chromatogr.* **264**, 63 (1983).
[17] J. Hermansson, *J. Chromatogr.* **325**, 375 (1985).
[18] T. Miwa, T. Miyakawa, M. Kayano and Y. Miyake, *J. Chromatogr.* **408**, 316 (1987).
[19] G. Blaschke, *J. Liquid Chromatogr.* **9**, 341 (1986).
[20] W. H. Pirkle, A. Tsipouras and T. J. Sowin, *J. Chromatogr.* **319**, 392 (1985).
[21] G. Vigh, G. Quintero and G. Farkas, *J. Chromatogr.* **506**, 481 (1990).
[22] D. W. Armstrong, A. M. Stalcup, M. L. Hilton, J. D. Duncan, J. R. Faulkner and S.-C. Chang, *Anal. Chem.* **62**, 1610 (1990).

14

Preparative direct chromatographic separation of enantiomers on chiral stationary phases

J. N. Kinkel, K. Cabrera and **F. Eisenbeiss**
E. Merck, Darmstadt, Germany

SUMMARY

Well-known optimization strategies of preparative chromatography have been applied for direct chromatographic HPLC separation of stereoisomers. General outlines are given for up-scaling procedures going from analytical direct to preparative-scale column dimensions and equipment.

The specific advantages of recycling chromatography techniques under overload conditions, especially for sample pairs with very small separation factors, have been demonstrated.

INTRODUCTION

By use of HPLC as a method for preparative isolation of compounds in laboratory-, technical- or production-scale, the process-determining parameters may be divided into three groups.

Chromatography parameters:
    selectivity (enantioselectivity), resolution, capacity factors, loadability of sorbent.

Solute parameters:
    solute concentration, solute volume, mass ratio of the compound of interest to by-products.

Hardware parameters:
    column dimensions, column efficiency, constructional details of the system, e.g. flow rate capability, pressure resistance, etc.

The results of a preparative chromatographic run are determined by the proper selection and combination of these various parameters and can be judged by the purity of the product, the throughput per unit time and the yield.

Depending on the specific needs, one or the other of these three properties has to be optimized and limits the economics of the total process.

In the chromatographic literature many theoretical and experimental papers have been published which describe strategies for optimization of up-scaling and demonstrate how to perform preparative chromatography in practice [1–5].

Therefore, to perform best in the preparative separation of enantiomers, knowledge and strategies of preparative chromatography, and knowledge of specific stereoselective interaction in the chromatogrphic separation of enantiomers on chiral stationary phase (CSP) have to be combined.

Indeed, the effects of load on the peak shape and peak position, as well as the influence on resolution in chiral separation, do not differ from those found in standard preparative chromatography, e.g. the effect of volume overload and concentration overload on peak shape and peak position (Figs 14.1–14.5). In practice, even concave isotherms for the enantiomers of a racemate have been observed (Fig. 14.2) [6,7].

GUIDELINE FOR THE OPTIMIZATION OF PREPARATIVE CHROMATOGRAPHIC SEPARATIONS OF ENANTIOMERS

Which chrial stationary phase?
Selection of CSP by:

— structural requirements [8,9,10]
— empirical
— use of database systems [11].

Which mobile phase?
Selection of mobile phase according to the requirements of preparative chromatography [12].

— good sample solubility, still maintaining high enantioselectivity
— short retention times
— recovery of solvent, price of solvent, environmental properties (disposal, toxicity), safety and purity.

Which sample amount?
Consequent overloading of the column.
— concentration overload
— volume overload
— form of adsorption/distribution isotherm
— use of strong solvents for sample dissolution.

Ch. 14] Preparative direct chromatographic separation

EXAMPLE 1: (R, S)-MANDELIC AMIDE

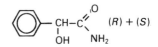

(a) Variation of sample concentration (analytical sample volume)

(c) Variation of sample volume (max. sample concentration)

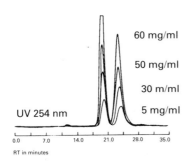

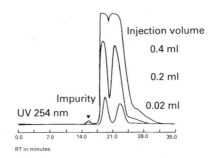

(b) Variation of sample volume (analytical sample concentration)

(d) Preparative separation (overload conditions)

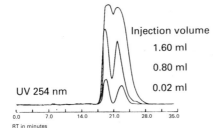

Fig. 14.1 — (R,S)-mandelic amide.

(a), (b), (c)
CSP: microcrystalline cellulose triacetate, 10 μm,
column: 250–10 mm
eluent: ethanol/methanol (70/30)
flow: 1.0 ml/min

(d)
column: 400–100 mm Prebar cartridge
particle size: 25–40 μm
flow: 100 ml/min

Which particle size and column dimensions?
With sufficient enantioselectivity, CSP-columns containing large-particle-size packings are preferred. Under overload conditions linear up-scale from analytical to preparative dimensions is possible.

268 Preparative direct chromatographic separation [Ch. 14

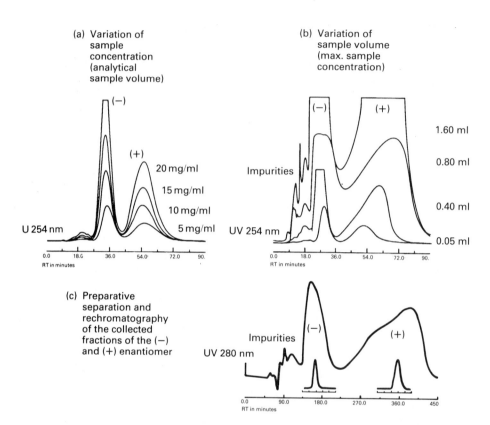

Fig. 14.2 — (R,S)-Troegers' base. This compound was the first enantiomer to be separated in a preparative scale on a CSP by Prelog and Wieland in 1941. Also, Troegers' base is the first example of an enantiomer to show different forms of adsorption isotherms on CTA.

(a), (b)
CSP   microcrystalline cellulose triacetate, 10 μm
column   250–10 mm
eluent   ethanol/water (90/10)
flow   1.0 ml/min
(c)
column   400–100 mm
particle size 25–40 μm
flow   100 ml/min

How to improve peak fractionation and economics (sample throughput and yield)?
— monitoring of concentration profiles of eluting peaks
— use of R- and S-configurations of CSP for preparative rechromatography of collected peak fractions according to peak shift under overload conditions and ratio of enantiomers

Preparative direct chromatographic separation

(a) Variation of load volume (analytical sample concentration)

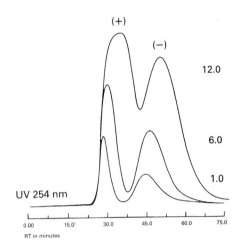

(b) Preparative separation (overload conditions)

(c) Analysis of the enantiomeric purity of collected fractions on a different type of CSP (ionically bonded DNBPG-column according to Pirkle)

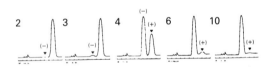

Fig. 14.3 — (R,S)-1-(9-Anthryl)-2,2,2-trifluorethanol.

(a)
CSP   microcrystalline cellulose triacetate, 10 µm,
column 250–4 mm
eluent ethanol
flow   1.0 ml/min
(b)
column   400–100 mm
particle size 25–40 µm
flow   100 ml/min

— use of different CSPs and/or both configurations of CSP for purity control
— use of recycling chromatographic techniques (closed-loop pumping approach, alternate pumping recycling) under overload conditions and peak-shaving in the case of closely eluting isomers.

In most cases, established analytical separation conditions have to be used to further elucidate the conditions to be applied for the preparative chromatography of

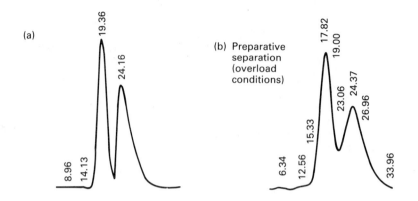

Fig. 14.4 — (rac.)-3-Benzyloxycarbonyl-2-*tert*-butyl-1-oxa-3-aza-cyclopentane-5-one [26].
(a)
CSP   Chiraspher®=poly-*N*-acronyl-(*S*)-phenylalanine ethyl ester bound to Lichrospher® diol, 5 µm.
column   250–4 mm
eluent   *n*-hexane/2-propanol (95/65)
flow   0.5 ml/min
sample   7 mg/ml
concentration
(b)
CSP   Chiraspher®, 25 µm
column   400–100 mm
flow   300 ml/min

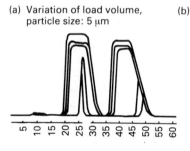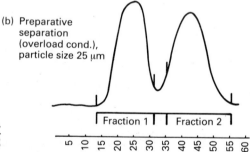

Fig. 14.5 — (rac.)-3-benzoyl-3-*tert*-butyl-1-oxa-3-aza-cyclopentane-5-one [26].
(a)
CSP   Chiraspher®, 5 µm
column   250–4 mm
eluent   *n*-hexane/2-propanol (98/2)
sample   0.5 mg/ml (95/5)
concentration
injection   0.005 ml, 0.5 ml, 2.5 ml, 3.0 ml, 3.5 ml
volume
flow   0.5 ml/min
(b)
CSP   Chiraspher®, 25 µm
column   400–100 mm
flow   310 ml/min

enantiomers. For selectivity factors larger than 1.4, the up-scaling stategies and the separation procedures for enantiomers of CSPs do not differ from the work with achiral substances and columns. On the contrary, enantiomeric samples offer several advantages:

— the enantiomers of a racemate behave similarly and predictably owing to a shift of k' values and because of solubility
— specific optical detection methods allow optimization of peak fractionation (use of polarimeter in combination with a UV-detector) [13,14].
— in most cases, the chemical purity of the racemic sample is high.

For use in preparative chromatography, the CSP should as stable as possible towards the necessary eluent. As the mobile phase selected often shows low solubility for the solute, the optimization for preparative use includes this parameter together with economic ones (Figs 14.6 and 14.7). To overcome the problem of low solubility of the

(a) Variation of sample volume
(analytical sample concentration; sample dissolved in eluent)

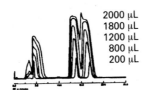

2000 μL
1800 μL
1200 μL
800 μL
200 μL

(b) Variation of sample concentration
(sample dissolved in strong eluent)

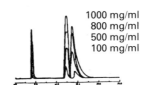

1000 mg/ml
800 mg/ml
500 mg/ml
100 mg/ml

(c) Variation of sample volume
(analytical sample concentration; sample dissolved in strong eluent)

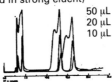

50 μL
20 μL
10 μL

Fig. 14.6 — (rac.)-2-Phenyl-1,3-oxathiolan-4-one.

(a)
CSP   Chiraspher®, 5 μm
column   250–4 mm
eluent n-hexane/dioxane (95/5)
flow   1.0 ml/min
sample concentration   0.5 mg/ml eluent
(b)
sample concentration   0.5 mg/ml in 100% dioxane
(c)
conditions as in (b)
injection volume   10 μm

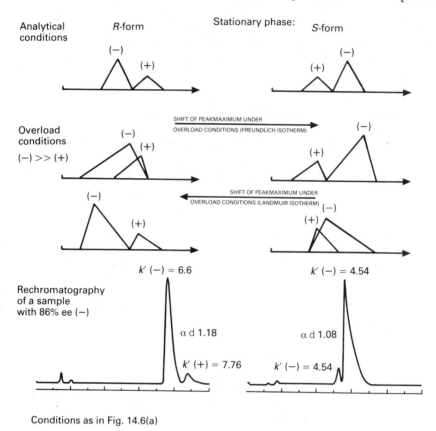

Fig. 14.7 — Schematical presentation of the influence of peak shift at overload conditions on the resolution of samples with large differences in enantiomeric composition.
Conditions as in Fig. 14.6(a)
CSP   Chiraspher®, 5 μm, R- and S-configuration of selector
column   250–25 mm
sample   1 g/ml in 100% dioxane (86% ee(-))
concentration
injection   0.8 ml
volume

sample in the mobile phase and to allow increased mass load on the column, the influence of strong solvents has been examined in several cases (e.g. Figs 14.8–14.15). Especially in the case of a relatively pure chemical composition of eanantiomers the chromatographic theory describing the effects on resolution for two adjacent peaks can be ideally used in the optimization of the separation. Based on the results of true concentration overload (Fig. 14.6), the self-displacement of enantiomers of 2-phenyl-1,3-oxathiolan-4-one in the rechromatography of a mixture containing a non-equal mass ratio was used advantageously in the optimization of the procedure as schematically and in practice is shown in Fig. 14.7. A preprequisite is the available CSPO in both configurations, R and S.

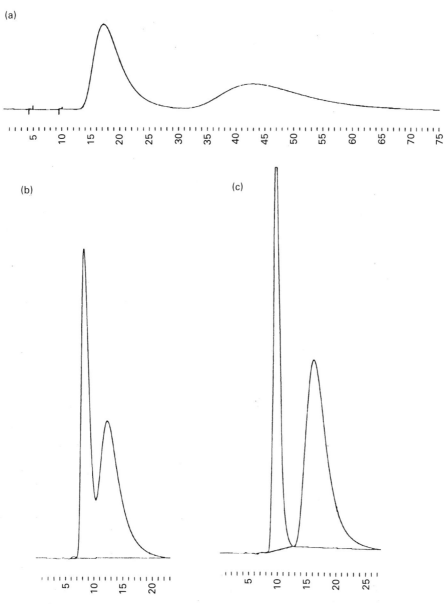

Fig. 14.8 — Influence of the mobile-phase composition on the separation of a platelet activating factor (PAF)-antagonist on microcrystalline cellulose triacetate.

(a)
solubility <140 mg/ml
$k'(+)$ 2.8
$k'(-)$ 8.65
α 3.0

(b) Ethanol/water (90/10)
solubility <100 mg/ml
$k'(+)$ 0.85
$k'(-)$ 1.73
α 2.03

(c) Methanol
solubility >1250 mg/ml
$k'(+)$ 1.13
$k'(-)$ 2.57
α 2.27

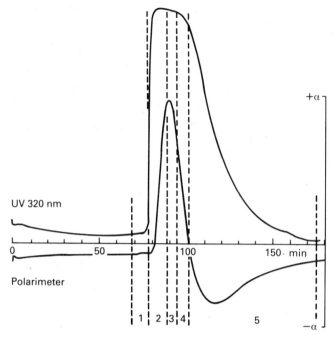

Fig. 14.9 — Preparative separation of a platelet activating factor (PAF)-antagonist with distomeric affinity.
CSP   microcrystalline cellulose triacetate, 25–40 µm
column   superformance glass column, 500–200 mm
sample   80 g racemic mixture in 80 ml methanol
flow   150 ml/min

Unfortunately, most of the CSPs are restricted to specific, solvents or solvent mixtures owing to a lack of stability or loss of enantioselectivity (Table 14.1). An optimization of selectivity by change of mobile phase is therefore limited and selectivity larger than 1.4 may not be achieved, Moreover, in the evaluation of pharmaceuticals, time plays an important role. In more than 98% of all cases, in a short period of time (several days to weeks) small amounts of pure enantiomers (mg to g) have to be provided in the laboratory for further pharmacological studies. In less than 1% of these cases, medium amounts of pure enantiomers (from 0 to several/ 100 g) have to be produced in a moderate period of time (weeks to month) on a technical scale. In an even smaller number of cases (less than 0.1%) the pure enantiomers are produced in large production scale (multi kg per year).

Preparative recycling chromatography
Since its first use back in 1962 by Porath and Bennich, recycling chromatography was regarded as a flexible method with high versatility to separate closely eluting compounds in the analytical and preparative scales [15–20]. Martin [21] made an extensive study on the parameters that effect the results of recycling as an analytical tool, while Coq and co-workers elucidated the limits of this technique in preparative

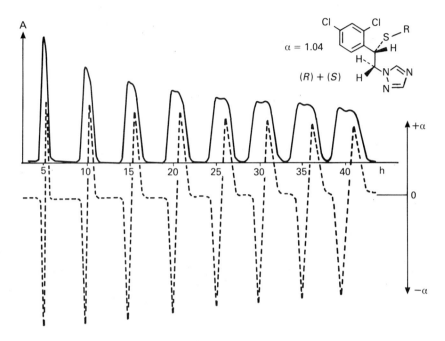

Fig. 14.10 — Separation of 2 g of a racemic fungicide by preparative recycling chromatography on microcrystalline cellulose triacetate [25].

LC [22]. The practical benefits of recycle-techniques, especially obtained in connection with peak-shaving, are higher throughput, lower separation time, lower solvent consumption and minimum column investment. Two basic systems are proposed to perform recycling chromatography [23,24]: (I) closed-loop recycling, and (II) alternate column recycling.

Closed-loop recycling is more convenient for preparative purposes, as it uses only one column and does not consume solvent during periods. Also, the experimental set-up of the instrumentation is quite simple and easy to derive from standard HPLC laboratory equipment, because the detector outlet from the column has only to be connected to the pump inlet via a multi-port selection valve. It is advantageous to reduce band-spreading, caused by extra-column effects like additional tubing or pumphead volume. To avoid cross-contamination, flushing out of the transport tubings must be possible. The system used in our experiments was assembled according to the details from Ref. [24].

Up to today, recycling chroamtography has not gained much popularity, even the results published since its first use were outstanding. For direct chromatographic enantiomer separation, Schlögl and co-workers evaluated microcrystalline cellulose triacetate with great success [23,24]. An example for the preparative separation of a racemic fungicide is given in Fig. 14.10 (taken with permission from Ref. [24]).

With CSPs based on high-efficiently HPLC-packings, remarkable achievements are possible. Fig. 14.11 gives the recycle chromatogram of j-nonalactone. This

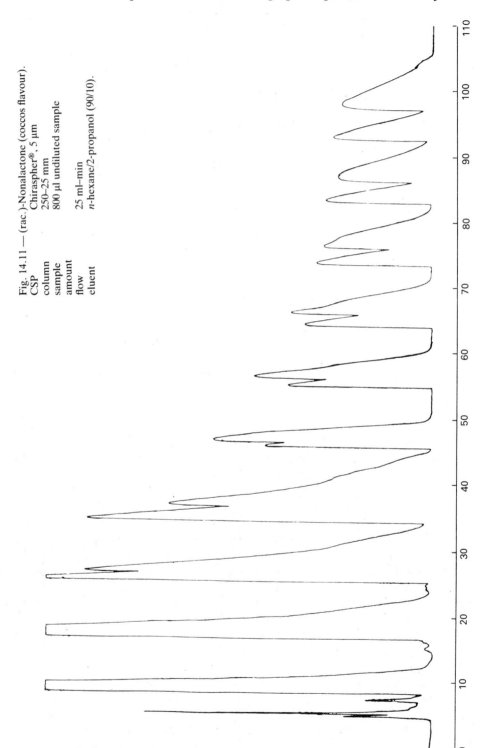

Fig. 14.11 — (rac.)-Nonalactone (coccos flavour).
CSP   Chiraspher®, 5 μm
column   250–25 mm
sample amount 800 μl undiluted sample
flow   25 ml-min
eluent   n-hexane/2-propanol (90/10).

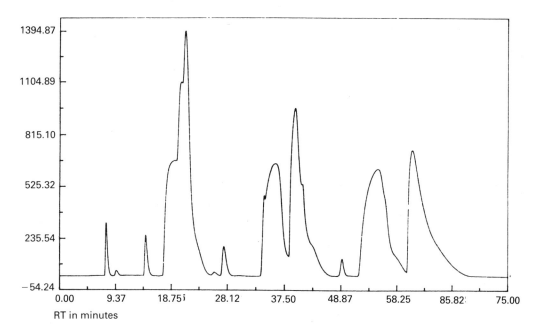

Fig. 14.12 — (rac.)-2-Phenyl-1,3-oxathiolan-4-one.

sample amount	1.5 g in 1.0 ml dioxan
CSP	Chiraspher®, 5 μm
column	250–25 mm
flow	25 ml/min
eluent	hexane/dioxan (95/5)

racemic sample is a liquid and was injected undiluted on to the CSP (Chiraspher®). Despite the fact, that under analytical conditions only a selectivity factor of 1.18 was reported, recycling under overload with peak shaving allowed the separation of more than 1000 μl pure sample in one run in less than 2 hours. The results of Fig. 14.8 have been combined with a recycling separation: 1.5 g of sample was dissolved in pure dioxane and was applied to the column (Fig. 14.12). Fig. 14.13 compares the results of a single-run separation of 5 g racemic 2-phenyl-$2H,4H$-1,3-dioxin-4-one performed on production-scale HPLC instrumentation with a 1.25 g separation by recycling on laboratory equipment. In both cases, the yield of the first eluting enantiomer was 0.6 g and a purity larger then 99% was achieved. Column dimensions, solvent consumption, yield and purity are compiled in Table 14.2. The heavy sample load on the 250–25 mm column is reduced on each successive pass by shaving the tailing edge of the second eluting enantiomer and therefore the column's effective separation efficiency increases. Compared with the lightly loaded system of the 400–100 column, in the former case only one fourth of the solvent consumption was needed, using only 1/25th of sorbent mass.

A fully automated run for the separation of 1 g of 1-benzoyl-2-*tert*-butyl-3-methyl-perhydropyrimidin-4-one is shown in Fig. 14.14 [25].

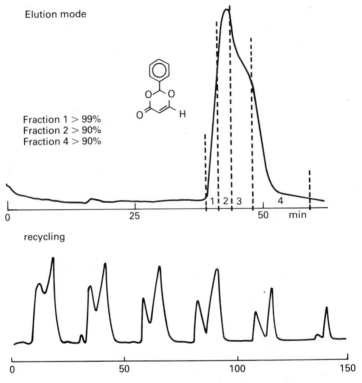

Fig. 14.13 — Comparison of a single column run with recycling chromatography for the separation of (rac.)-2-phenyl-$2H$,$4H$-1,3-dioxin-4-one.
(a) single column run,
(b) recycling operation (conditions given in Table 14.2).

The analysis of enantiomeric purity revealed 098% for the first enantiomers and 90% for the second eluting enantiomer. Sometimes, sample dissolution is not possible with common solvents, as in the case of the 3,5-dinitrobenzoyl-β-amino acid of Fig. 14.15, when 200 mg racemate, dissolved in 0.5 ml pyridin, were successfully separated in eight cycles.

CONCLUSION

Preparative direct chromatographic separation of enantiomers becomes easy when the main parameter, the selection of an appropriate CSP showing some enantioselectivity towards the racemic sample has been solved. Time-consuming optimization of enantioselectivity or the expensive evaluation of further CSPs may be unnecessary when closed-loop recycling chromatography and peak shaving is applied.

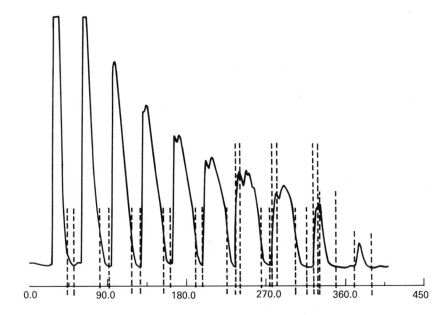

Fig. 14.14 — (rac.)-1-Benzoyl-2-*tert*-butyl-3-methyl-perhydropyrimidin-4-one.

CSP	Chiraspher®, 5 µm
column	250–50 mm
eluent	hexane/2-propanol (97/3)
flow	80 ml/min
sample amount	1 g dissolved in 2 ml 2-propanol

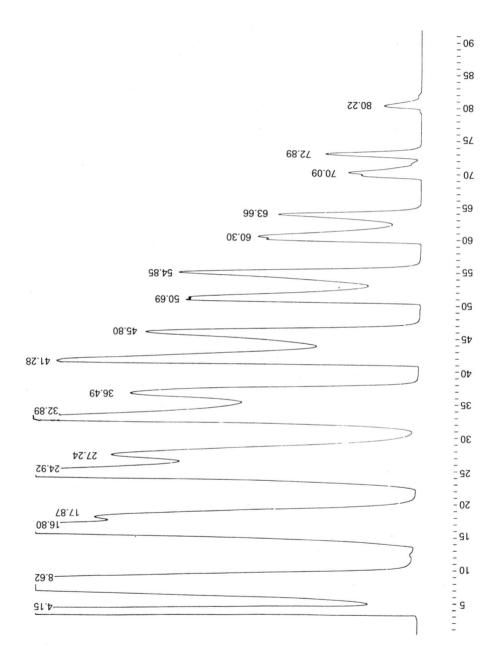

Fig. 14.15 — (rac.)-3,5-Dinitrobenzoyl-β-phenyl leucine.
CSP Poly-*n*-acryloyl-*S*-phenylalanine diethyl amide bonded to Lichrospher diol, 5 μm (research sample)
sample amount 200 mg dissolved in 0.5 ml pyridin
flow 39 ml/min

Table 14.1 — Chiral sorbents and possible eluent systems

Type of ligand and classification of sorbent	Typical eluent composition (v/v)	Comments
Sorbents for exclusive use in adsorption mode		
Ionically bonded π-acceptor and π-donors	Hexane/2-propanol 90/10	Up to 20% 2-propanol
Amino acid amides	Hexane/2-propanol	
Monosaccharides and derivatives bound to silica	Hexane/2-propanol/ dicloromethane	
Sorbents for exclusive use in reversed-phase mode		
Cyclodextrins	Water/org. solvent	Up to 80% org. solvent
Proteins	Aqueous buffered solutions pH 5–9	Up to 5% org. modifier
Chelate-forming agents (Ligand exchange)	0.25 mm Cu(II) sulphate pH 4–5	Org. modifiers possible
Crown ethers	Strong inorganic acids (10^{-5}–10^{-3}M)	
Sorbents for use in adsorption and reversed-phase mode (main application area is marked by I the above)		
Microcrystalline Cellulose tri-esters and tri-carbamates(II)	Hexane/ethanol 90/10 Methanol Ethanol/water	Limitations for solvents like chlorinated hydrocarbons, Ketones, esters
Amorphous cellulose- and other polysaccarid- derivatives (I,II) (Carbamates and esters)	Hexane/2-propanol Water/methanol Ethanol Methanol	Limitations for polar solvents and chlorinated hydrocarbons
Poly)triphenylmethyl-methancrylate) I (II)	Hexane/2-propanol	Limitations for solvents like aromatic hydocarbons, chloroform, THF
Bonded π-acceptors and π-donors (Pirkle) I (II)	Hexane/2-propanol Water/alcohol	
Polyacrylamides (Chiraspher) I, II	Hexane/dioxane Hexane/dichloromethane Water/org. solvents	

Table 14.2 — Comparison or recycling mode with single-column operation ((rac.)-2-phenyl-2H,4H-1,3-dioxin-4-one). CSP: Chiraspher®, 5 µm, resp. 25 µm

	Normal mode	Recycling
Sample amount: (in dioxane)	5 g	1.25 g
Column dimension	400–100 mm	250–25 mm
Amount of stationary	1700 g	80 g
Flow	50 ml/min	39 ml/min
Particle size	25 µm	5 µm
Solvent consumption	3800 ml	1000 ml
Yield 1. enantiomer (ee>99%)	0.6 g	0.6 g
Yield 2. enantiomer	0.2 g	0.6 g
Recovery	?	>95%
Cycle time	75 min	120 min

REFERENCES

[1] F. Eisenbeiß, S. Ehlerding, A. Wehrli and J. F. K. Huber, *Chromatographia*, **23**, 657 (1985).
[2] S. Ghodbane and G. Guiochon, *J. Chromatogr.*, **444**, 275 (1988).
[3] K. Hostettmann, M. Hostettmann and A. Marston, *Preparative Chromatography Techniques*, Springer-Verlag, Berlin (1986).
[4] W. H. Pirkle and B. C. Hamper, in *Preparative Liquid Chromatography, J. Chromatogr. Library*, **38**, 235–287 (1987).
[5] S. Jacobson, S. Golshan-Shirazi and G. Guichon, *J. Am. Chem. Soc.*, **112**, 6492–6498 (1990).
[6] J. Kinkel, K. Reichert and P. Knöll, *GIT, Supplement Chromatogr.*, 104–112.
[7] I. Isaksson, H. Wennerström and O. Wennerström, *Tetrahedron*, **44**, 1697 (1988).
[8] I. Wainer: *A Practical Guide to the Selection of Chiral Stationary Phases*, J. T. Baker (ed.), Philipsburg, NJ (1988).
[9] M. Zief and L. J. Crane, *Chromatographic Chiral Separations Chromatographic Sci. Series*, **40**, M. Dekker, NY (1988).
[10] S. G. Allenmark, *Chromatographic Enantioseparation: Methods and Applications*, (R. A. Chalmers, ed.) Ellis Horwood, Chichester (1988).
[11] For example, C. Roussel and P. Piras, 'CHIRBASE', a graphic database for the enantiomeric resolution of racemic mixtures by chiral liquid chromatography, ENSSPICAM Marseille (1990).
[12] P. D. McDonald and B. A. Bidlingmeyer in *Preparative Liquid Chromatography, J. Chromatogr. Library*, **38**, 66–81 (1987).
[13] A. Mannschreck, M. Mintas, G. Becher and G. Stühler, *Angew. Chem.*, **92**, 490 (1980).

[14] A. F. Drake, J. M. Gould and S. F. Mason, *J. Chromatogr.*, **202**, 239–245 (1980).
[15] J. Porath and H. Bennich, *Arch. Biochem. Biophys. Suppl.*, **1**, 152–156 (1962).
[16] K. J. Bombaugh and R. F. Levangie, *J. Chromatogr. Sci*, **8**, 560–566 (1970).
[17] S. van der Wal, *Chromatographia*, **22**, 81–87 (1986).
[18] H. Kalaz, J. Nagy and J. Knoll, *J. Chromatogr.* **105**, 35 (1975).
[19] J. Leseca and C. Quivoron, *Analysis*, **4**, 120 (1976).
[20] E. Farmakalidis and P. A. Murphy, *J. Chromatogr.*, **316**, 510 (1984).
[21] M. Martin, F. Verilon, C. Econ and G. Guiochon, *J. Chromatogr.* **4**, 237–249 (1976).
[22] B. Coq, G. Cretier, J. L. Rocca and J. Vialle, *J. Liquid Chromatogr.*, **4**, 237–249 (1981).
[23] K. Schlögl and M. Widhalm, *Monatsh. Chem.*, **115**, 1113–1120 (1984).
[24] A. Werner, *Kontakte (Darmstadt)* (3) 50–56 (1989).
[25] E. Juaristi, D. Qintana, B. Lamatsch and D. Seebach, *J. Org. Chem.*, **56**, 2553–2557 (1991).
[26] D. Seebach, S. G. Müller, U. Gysel and J. Zimmermann, *Hel. Chim. Acta*, **71**, 1303–1318 (1988).

Index

adsorption isotherm, 177
adsorption kinetics, 150
adsorption process, 176
   design procedures, 176
adsorption, specific v. non-specific, 237
adsorption techniques, 37
affinity chromatography, 227
affinity ligands
   mimetic, 244
   natural v. synthetic, 241
   separation mechanism, 239
   textile dyes, 242
affinity tags, 232
air sensors, 50
analytical HPLC, 9
   definition, 10
analytical and preparative LC
   comparison, 20
   equipment comparison, 11
automations, 52, 144
axial compression, 91

backflushing, 29
basic component of chromatogaphy system, 39
batch technique for anion exchange, 154
bed stability, 86
binary fractionation, 210

capital cost, 144
celluloses
   anion exchange, 149
   microcrystalline triacetate, 257, 261, 267–269, 273, 274
chambers, 167
chirality, 250
chiral separation, analytical, 255
chiral stationary phases
   classification, 255
   Type I, 255
   Type II, 256
   Type III, 256
   Type IV, 259
   Type V, 260
chiraspher, 270–272, 276, 277, 279
chromatography

affinity, 227
analytical, 9
continuous co-current, 207
continuous counter current, 207
gel-filtration, 229
hydrophobic interaction, 228
ion-exchange, 226
reverse phase, 105
size exclusion, 105
chromatographic parameters, 265
column
   cartridges, 89
   choice, 230
   compression, 89
   dimensions, 267
   empty, 88
   end fitting, 83
   large size, 83
   length, 159
   life, 139
   loading capacity, 23
   maintenance, 230
   packing, 134, 137
   prepacked, 88
   regular type, 88
   size variation of peak volume, 34
   size, weight of packing, 23
   wall thickness, 83
column technique for anion exchange, 154
comparison: multistep v. affinity, 238
component
   column, 45
   of plants, 59
   pumps, 46
   selection, 43
components, 63
composition of material, 27
compression
   axial, 91
   column, 89
   radial, 90
   static and dynamic, 89
computer-based system, 53
conductivity, 51
continuous co-current chromatography, 207

Index

continuous counter current chromatography, 209
continuous moving bed contactor, 162
control of plant requirement, 66
controller
    dedicated, 52
    programmed logic, 52
crown ethers, 259
cyclodextrin, 258

de-aerated fluid, 217
dedicated controller, 52
design
    chromatographic media, 69
    diaphragm monitoring pump, 61
    diaphragm valve, 61
    plant for prep LC, 57
design factors for prep media, 70
desorption kinetics of media, 150
desorption technique, 37
detection, 140
detector, UV, 118
diastereoisomer, 111, 114

economic consideration, 16
economic comparison, 173
economic factors, 15
effluent profiles, 217
electrical safety requirement, 42
eluent
    feed pump, 136
    filtration, 135
    mode, 134
    mixing, 138
    preparation, 134
    properties, 64
    pumps, 64
    reproducibility, 134
    system design, 135
eluent regime
    gradient, 121
    isocratic, 121
elution and sample preconcentration, 34
elution tests, 185
elution time v. particle size, 18
end fitting, 83
equipment constraint, 26
explosion proofing, 43

filters, 49
flow distribution, 84
flowrate, 48
flowrate v. particle size, 17
fraction collection, 142
frontal elution, 38

gel filtration chromatography, 229
gradient, 49
gradient eluent regimes, 127
gradient-shallow, 127

hardware consideration, 124–127
hydrophobic interaction, 241

hydrophobic interaction chromatography, 228
hygiene, 43, 49

injection technique, 14
instrument parameters
    analytical, 12
    preparative, 12
instrumentation, 51
ion-exchange, 105
ion-exchange chromatography, 105
ISEP, 162
    as chromatographic separator, 207

labetalol, 254
laboratory simulation, 189
lactic acid purification, 192–196
loading capacity v. column size, 23
loading v. particle size, 18
lysine, 175

mandelic amide, 267
maintenance of column, 236
material compatibility, 41
matrices, 225
matrix, 240
maximum capacity, 156
media
    adsorption kinetics, 150
    availability, 148
    capacity, 149
    desorption kinetics, 150
    quality, 148
    regeneration and sanitization, 148
    reuse, 159
    sanitization, 150
    selectivity, 149
    throughput, 148
methods of feed addition, 217
microcrystalline cellulose triacetate, 257, 261, 267–269, 273, 274
mixing and degassing, 122
mobile phase, 25
    consideration, 106
    diffusivity coefficient, 26

nonalactone, 276
non-process preparative separations, 96

packing and unpacking process, 84
packing materials
    analytical, 12
    chemical properties, 28
    physical properties, 27
    preparative, 12
particle shape, 78, 94
    availability, 79
    cost, 79
    ease of packing, 80
    quality, 80
    range, 80
    surface chemistry, 79
particle size, 71, 79, 226

cost of packing, 72
   distribution, 93
   ease of packing, 71
   loadability, 72
pipework, 50
pirkle-type, phases, 262
plant concept, 58
plant control, 64
plant design for prep LC, 57
plate number, 15
pore diameter, 27
pore size, 74
pore volume, 74
premixing, 122
preparative recycling chromatography, 274
pressure, 41, 51
process chromatography, 97
   strategy for optimization, 98
process design, 229
product purity, 14, 93
product recovery, 142
programmed logic controller, 52
protein engineering, 231

racemic fungicide, 275
recycle, 35
regeneration of media, 148
regular-type column, 88
reliability, 45
resin
   evaluation, 191
   loading, 216
   preconditioning, 216
   selection, 189
reverse phase, 105
reverse phase eluents, 108

sample amount, 266
sample compatability, 49
sample loading, 136
sample preconcentration, 34
sample preparation, 136
sample stability, analytical and preparative, 14
scale up, 15, 92, 230, 260
separation of enantiomers
   direct preparative, 265
   optimization, 266
serviceability, 45
silica, 105
solute parameters, 265
solvent recovery, 143
sorbent selection, 189
specific v. non-specific adsorption, 237
stationary phase consideration, 104
straight phase eluent, 108
strength of material, 27
surface area, 27
synthetic polymers, 259
system design, 45
system operation, 13
system requirement, 40

technique, valve, 36
technology column, 87
throughput, 25
   media, 148
touching bands, 29–31
Troeger's base, 268
Type I chiral stationary phase, 255
Type II chiral stationary phase, 256
Type III chiral stationary phase, 256
Type IV chiral stationary phase, 259
Type V chiral stationary phase, 260

ELLIS HORWOOD SERIES IN
APPLIED SCIENCE AND INDUSTRIAL TECHNOLOGY

Series Editor: Dr D. H. SHARP, OBE, former General Secretary, Society of Chemical Industry; formerly General Secretary, Institution of Chemical Engineers; and former Technical Director, Confederation of British Industry.

MECHANICS OF WOOL STRUCTURES
R. POSTLE, University of New South Wales, Sydney, Australia, G. A. CARNABY, Wool Research Organization of New Zealand, Lincoln, New Zealand, and S. de JONG, CSIRO, New South Wales, Australia

MICROCOMPUTERS IN THE PROCESS INDUSTRY
E. R. ROBINSON, Head of Chemical Engineering, North East London Polytechnic

BIOPROTEIN MANUFACTURE: A Critical Assessment
D. H. SHARP, OBE, former General Secretary, Society of Chemical Industry; formerly General Secretary, Institution of Chemical Engineers; and former Technical Director, Confederation of British Industry

QUALITY ASSURANCE: The Route to Efficiency and Competitiveness, Second Edition
L. STEBBING, Quality Management Consultant

QUALITY MANAGEMENT IN THE SERVICE INDUSTRY
L. STEBBING, Quality Management Consultant

INDUSTRIAL CHEMISTRY
E. STOCCHI, Milan, with additions by K. A. K. LOTT and E. L. SHORT, Brunel

REFRACTORIES TECHNOLOGY
C. STOREY, Consultant, Durham; former General Manager, Refractories, British Steel Corporation

COATINGS AND SURFACE TREATMENT FOR CORROSION AND WEAR RESISTANCE
K. N. STRAFFORD and P. K. DATTA, School of Material Engineering, Newcastle upon Tyne Polytechnic, and C. G. GOOGAN, Global Corrosion Consultants Limited, Telford

TEXTILE OBJECTIVE MEASUREMENT AND AUTOMATION IN GARMENT MANUFACTURE
G. STYLIOS, Department of Industrial Technology, University of Bradford

PREPARATIVE AND PROCESS-SCALE LIQUID CHROMATOGRAPHY
G. SUBRAMANIAN, Department of Chemical Engineering, Loughborough University of Technology

INDUSTRIAL PAINT FINISHING TECHNIQUES AND PROCESSES
G. F. TANK, Educational Services, Graco Robotics Inc., Michigan, USA

MODERN BATTERY TECHNOLOGY
Editor: C. D. S. TUCK, Alcan International Ltd, Oxon

FIRE AND EXPLOSION PROTECTION: A Systems Approach
D. TUHTAR, Institute of Fire and Explosion Protection, Yugoslavia

PERFUMERY TECHNOLOGY 2nd Edition
F. V. WELLS, Consultant Perfumer and former Editor of *Soap, Perfumery and Cosmetics,* and
M. BILLOT, former Chief Perfumer to Houbigant-Cheramy, Paris, Président d'Honneur de la Société Technique des Parfumeurs de la France

THE MANUFACTURE OF SOAPS, OTHER DETERGENTS AND GLYCERINE
E. WOOLLATT, Consultant, formerly Unilever plc

Current themes in diabetes care

Papers based on the Interfaces in Medicine Conference held jointly by the Royal College of Physicians, London

Edited by
Ian G Lewin MD FRCP
*Consultant Physician, North Devon District Hospital
Barnstaple, Devon*
and
Carol A Seymour PhD FRCP
*Professor of Clinical Biochemistry and Metabolism
St George's Hospital Medical School, London*

1992

ROYAL COLLEGE OF PHYSICIANS OF LONDON

Acknowledgement

The Royal College of Physicians is grateful to Bayer Diagnostics (UK) Ltd for their assistance in the production of this book.

Royal College of Physicians
11 St Andrews Place, London NW1 4LE

© 1992 Royal College of Physicians of London
ISBN 1 873240 29 5

Typeset by Oxprint Ltd, Aristotle Lane, Oxford OX2 6TR
Printed by Bayer Diagnostics (UK) Ltd, Evans House, Hamilton Close, Houndmills, Basingstoke, Hants RG21 2YE